Bee Conservation
Evidence for the effects of interventions

Lynn V. Dicks, David A. Showler
& William J. Sutherland

Synopses of Conservation Evidence, Volume 1

Pelagic Publishing | www.pelagicpublishing.com

Published by **Pelagic Publishing**
www.pelagicpublishing.com
PO Box 725, Exeter, EX1 9QU

Bee Conservation
Evidence for the effects of interventions
Synopses of Conservation Evidence, Volume 1

ISBN 978-1-907807-00-8 (Pbk)
ISBN 978-1-907807-01-5 (Hbk)

British Library Cataloguing in Publication Data
A catalogue record for this book is available from the British Library.

Contents

Advisory board

We thank the following people for advising on the scope and content of this synopsis.

Professor **Andrew Bourke**, University of East Anglia, UK
Dr Claire Carvell, Centre for Ecology and Hydrology, UK
Mike Edwards, Bees, Wasps and Ants Recording Society, UK
Professor **Dave Goulson**, University of Stirling
  & Bumblebee Conservation Trust, UK
Dr Claire Kremen, University of California, Berkeley, USA
Dr Peter Kwapong, International Stingless Bee Centre,
  University of Cape Coast, Ghana
Professor **Ben Oldroyd**, University of Sydney, Australia
Dr Juliet Osborne, Rothamsted Research, UK
Dr Simon Potts, University of Reading, UK
Matt Shardlow, Director, Buglife, UK
Dr David Sheppard, Natural England, UK
Dr Nick Sotherton, Game and Wildlife Conservation Trust, UK
Professor **Teja Tscharntke**, Georg-August University, Göttingen, Germany
Mace Vaughan, Pollinator Program Director, The Xerces Society, USA
Sven Vrdoljak, University of Stellenbosch, South Africa
Dr Paul Williams, Natural History Museum, London, UK

About the authors

Lynn Dicks is a Research Associate in the Department of Zoology, University of Cambridge.

David Showler is a Research Associate in the School of Biological Sciences, University of East Anglia and the Department of Zoology, University of Cambridge.

William Sutherland is the Miriam Rothschild Professor of Conservation Biology at the University of Cambridge.

Acknowledgements

This synopsis was prepared with funding from Arcadia. The Conservation Evidence project has also received funding from the Natural Environment Research Council (NERC) and the British Ecological Society (BES).

We also thank Dr Barbara Gemmill-Herren, Dr Rob Pople and Dr Stephanie Prior for their help and advice.

1. Introduction

1.1 The purpose of Conservation Evidence synopses

This book, *Bee Conservation*, is the first in a series of synopses that will cover different species groups and habitats, gradually building into a comprehensive summary of evidence on the effects of conservation interventions for all biodiversity throughout the world.

By making evidence accessible in this way, we hope to enable a change in the practice of conservation, so it can become more evidence-based. We also aim to highlight where there are gaps in knowledge.

1.2 Who this synopsis is for

If you are reading this, we hope you are someone who has to make decisions about how best to support or conserve biodiversity. You might be a land manager, a conservationist in the public or private sector, a farmer, a campaigner, an advisor or consultant, a policymaker, a researcher or someone taking action to protect your own local wildlife. Our synopses summarise scientific evidence relevant to your conservation objectives and the actions you could take to achieve them.

We do not aim to make your decisions for you, but to support your decision-making by telling you what evidence there is (or isn't) about the effects that your planned actions could have.

When decisions have to be made with particularly important consequences, we recommend carrying out a systematic review, as the latter is likely to be more comprehensive than the summary of evidence presented here. Guidance on how to carry out systematic reviews can be found from the Centre for Evidence-Based Conservation at the University of Wales, Bangor (www.cebc.bangor.ac.uk).

Table 1.1 The Conservation Evidence synopses concept

Conservation Evidence synopses do	Conservation Evidence synopses do not
• Bring together scientific evidence captured by the Conservation Evidence project (over 2,000 studies so far) on the effects of interventions to conserve wildlife	• Include evidence on the basic ecology of species or habitats, or threats to them
• List all realistic interventions for the species group or habitat in question, regardless of how much evidence for their effects is available	• Make any attempt to weight or prioritise interventions according to their importance or the size of their effects
• Describe each piece of evidence, including methods, as clearly as possible, allowing readers to assess the quality of evidence	• Weight or numerically evaluate the evidence according to its quality
• Work in partnership with conservation practitioners, policymakers and scientists to develop the list of interventions and ensure we have covered the most important literature	• Provide answers to conservation problems. We provide scientific information to help with decision-making

1.3 The Conservation Evidence project

The Conservation Evidence project has three parts:

1. An online, **open access journal** *Conservation Evidence* publishes new pieces of research on the effects of conservation management interventions. All our papers are written by, or in conjunction with, those who carried out the conservation work and include some monitoring of its effects.

2. An ever-expanding **database of summaries** of previously published scientific papers, reports, reviews or systematic reviews that document the effects of interventions.

3. **Synopses** of the evidence captured in parts one and two on particular species groups or habitats. Synopses bring together the evidence for each possible intervention. They are freely available online and available to purchase in printed book form.

These resources currently comprise over 2,000 pieces of evidence, all available in a searchable database on the website www.conservationevidence.com.

Alongside this project, the Centre for Evidence-Based Conservation (www.cebc.bangor.ac.uk) and the Collaboration for Environmental Evidence (www.environmentalevidence.org) carry out and compile systematic reviews of evidence on the effectiveness of particular conservation interventions. These systematic reviews are included on the Conservation Evidence database.

Of the 59 bee conservation interventions identified in this synopsis, one is the subject of a current systematic review (Systematic Review number 72: Does delaying the first mowing date increase biodiversity in European farmland meadows? www.environmentalevidence.org/SR72.html).

We identify an immediate need for a systematic review in relation to one other set of interventions (agri-environment schemes), and a potential need for systematic reviews for three further interventions, should they become more widely practised (nest boxes for solitary bees and captive rearing of bumblebees or solitary bees).

1.4 Scope of the Bee Conservation synopsis

This synopsis covers evidence for the effects of conservation interventions for native, wild bees.

It is restricted to evidence captured on the website www.conservationevidence.com. It includes papers published in the journal *Conservation Evidence*, evidence summarised on our database and systematic reviews collated by the Collaboration for Environmental Evidence.

It does not include evidence from the substantial literature on husbandry methods for the largely domesticated honey bee *Apis mellifera*. It does include husbandry methods where they are relevant to native, wild bee species that are declining or threatened, such as bumblebees (*Bombus* spp.) and stingless bees (Meliponinae). Although the number of managed honey bee colonies is known to have declined in Europe and America, it is seldom the native subspecies that is kept and so we consider this to be outside the remit of Conservation Evidence. We do include some interventions and evidence relating to the conservation of subspecies of *Apis mellifera* in areas where they are native.

Evidence from all around the world is included. If there appears to be a bias towards evidence from northern European or North American temperate environments, this reflects a current bias in the published research that is available to us.

1.5 How we decided which bee conservation interventions to include

Our list of interventions has been agreed in partnership with an Advisory Board made up of international conservationists and academics with expertise in bee conservation. Although the list of interventions may not be exhaustive, we have tried to include all actions that have been carried out or advised to support populations or communities of wild bees.

1.6 How we reviewed the literature

In addition to evidence already captured by the Conservation Evidence project, we have searched the following sources for evidence relating to bee conservation: four specialist bee or insect conservation journals, from their first publication date to the end of 2009 (*Apidologie, Journal of Apicultural Research, Insect Conservation and Diversity, Journal of Insect Conservation*); ISI

Web of Knowledge searched for papers with 'bee' as a search term, from 1997 to 2009 inclusive; all reports concerning bees published by Natural England or the UK Bumblebee Working Group up to 2009; other relevant papers or books frequently cited within the bee conservation literature, going back to 1912.

In total, 168 individual studies are covered in this synopsis, all included in full or in summary on the Conservation Evidence website.

The criteria for inclusion of studies in the Conservation Evidence database are as follows:

- There must have been an intervention that conservationists would do

- Its effects must have been monitored quantitatively

In some cases, where a body of literature has strong implications for conservation of a particular species group or habitat, although it does not directly test interventions for their effects, we refer the reader to this literature. For example, the proportion of natural habitat in farmland has often been shown to affect bee diversity, but no studies have yet intervened by restoring natural or semi-natural habitat and monitoring the effect on bees in surrounding farmland. In cases such as these, we briefly refer to the relevant literature, but present no evidence.

1.7 How the evidence is summarised

Conservation interventions are grouped primarily according to the relevant direct threats, as defined in the International Union for the Conservation of Nature (IUCN)'s Unified Classification of Direct Threats (www.iucnredlist.org/technical-documents/classification-schemes). In most cases, it is clear which main threat a particular intervention is meant to alleviate or counteract. Interventions to help bees threatened by agricultural land use change are very different from those intended to avoid the adverse effects of invasive species, for example.

Not all IUCN threat types are included, only those that threaten bees, and for which realistic conservation interventions have been suggested.

We have separated out three important categories of conservation action, as defined by the IUCN, which are relevant to a variety of situations, habitats and threats. They are: Chapter 10 *Providing artificial nest sites*, Chapter 11 *Captive breeding and rearing of wild bees* and Chapter 12 *Education and awareness-raising*. These respectively match the following categories of conservation actions defined by the IUCN: 'species management: species recovery', 'species management: *ex situ* conservation' and 'education and awareness'.

Normally, no intervention or piece of evidence is listed in more than one place, and when there is ambiguity about where a particular intervention should fall there is clear cross-referencing. The only exception to this is in Chapter 4.1 *Introduce agri-environment schemes that reduce spraying*. Due to the prevalence of review papers in this section, some individual studies are referred to that are also referred to in more specific sections on particular agri-environment prescriptions.

In the text of each section, studies are presented in chronological order, so the most recent evidence is presented at the end. The summary text at the start of each section groups studies according to their findings.

At the start of each chapter, a series of **key messages** provides a rapid overview of the evidence. These messages are condensed from the summary text for each intervention.

In general, we do not update taxonomy, but employ species names used in the original paper. However, in some cases it is sensible to replace the names with their modern equivalent. For example, papers from the early 20th century may describe bumblebees in the genus *Bremus* not *Bombus*. This would be changed and *Bremus* included as a keyword in the database of summaries. Any replacement names are those used in the ITIS (Interagency Taxonomic Information System) World Bee Checklist (www.itis.gov/beechecklist. html). Where possible, common names and Latin names are both given the first time each species is mentioned within each intervention.

Background information is provided where we feel recent knowledge is required to interpret the evidence. This is presented separately and relevant references included in the reference list at the end of each intervention section.

References containing evidence of the effects of interventions are marked with a weblink icon (www↗). In electronic versions of the synopsis, they are hyperlinked directly to the Conservation Evidence summary. If you do not have access to the electronic version of the synopsis, typing the first author's name into the 'Quick Search' facility on www.conservationevidence. com is the quickest way to locate summaries.

The information in this synopsis is available in three ways:

As a book, published by Pelagic Publishing and for sale from:
www.pelagicpublishing.com

As a .pdf file to download from:
www.conservationevidence.com

As text for individual interventions on the searchable database at:
www.conservationevidence.com

1.8 Terminology used to describe evidence

Unlike systematic reviews of particular conservation questions, we do not quantitatively assess the evidence, or weight it according to quality. However, to allow you to interpret evidence, we make the size and design of each trial we report clear. The table below defines the terms that we have used to do this.

The strongest evidence comes from randomised, replicated, controlled trials with paired sites and before and after monitoring.

Table 1.2 Terminology used to describe evidence in the Conservation Evidence synopses

Term	Meaning
Site comparison	A study that considers the effects of interventions by comparing sites that have historically had different interventions or levels of intervention.
Replicated	The intervention was repeated on more than one individual or site. In conservation and ecology, the number of replicates is much smaller than it would be for medical trials (when thousands of individuals are often tested). If the replicates are sites, pragmatism dictates that between five and ten replicates is a reasonable amount of replication, although more would be preferable. We provide the number of replicates wherever possible, and describe a replicated trial as 'small' if the number of replicates is small relative to similar studies of its kind.
Controlled	Individuals or sites treated with the intervention are compared with control individuals or sites not treated with the intervention.
Paired sites	Sites are considered in pairs, within which one was treated with the intervention and the other was not. Pairs of sites are selected with similar environmental conditions, such as soil type or surrounding landscape. This approach aims to reduce environmental variation and make it easier to detect a true effect of the intervention.
Randomised	The intervention was allocated randomly to individuals or sites. This means that the initial condition of those given the intervention is less likely to bias the outcome.
Before-and-after trial	Monitoring of effects was carried out before and after the intervention was imposed.
Review	A conventional review of literature. Generally, these have not used an agreed search protocol or quantitative assessments of the evidence.
Systematic review	A systematic review follows an agreed set of methods for identifying studies and carrying out a formal 'meta-analysis'. It will weight or evaluate studies according to the strength of evidence they offer, based on the size of each study and the rigour of its design. All environmental systematic reviews are available at: www.environmentalevidence.org/index.htm

1.9 How you can help to change conservation practice

If you know of evidence relating to bee conservation that is not included in this synopsis, we invite you to contact us, via the www.conservationevidence.com website.

Following guidelines provided on the site, you can submit a summary of a previously published study, or submit a paper describing new evidence to the *Conservation Evidence* journal. We particularly welcome summaries written by the authors of papers published elsewhere, and papers submitted by conservation practitioners.

2. Threat: residential and commercial development

Key Messages

Plant parks and gardens with appropriate flowers

Three North American trials have found more wild bees in gardens planted with bee forage plants, either relative to conventionally managed gardens or following planting.

Practise 'wildlife gardening'

A UK site comparison study found more species of bumblebee in domestic city gardens with lower intensity of management, a measure reflecting tidiness of the garden and use of pesticides.

Protect brownfield sites

We have captured no evidence for the effects of interventions to protect brownfield sites from insensitive re-development.

Conserve old buildings or structures as nesting sites for bees

We have captured no evidence for the effects of conserving old buildings and structures suitable for nesting, or containing nesting wild bees.

For all evidence relating to the use of **nest boxes**, see Chapter 10 *Providing artificial nest sites for bees*.

2.1 Plant parks and gardens with appropriate flowers

- Two replicated trials in the USA and Canada have found more wild bees (either more species or more individuals) in gardens planted with bee forage or native plants, relative to conventionally managed gardens. Another USA trial found more bee species after the addition of bee forage plants to a community garden. Three trials in the UK or USA have shown that native flowering plants or bee forage plants are well used by wild bees when planted in gardens. A UK trial demonstrated that some popular non-native or horticulturally modified garden flowers are not frequently visited by insects, despite providing nectar in some cases.

Natural shaped, rather than horticulturally modified varieties of garden plants are recommended for foraging insects. A trial of nearly natural and horticulturally modified varieties of six popular garden plants in the Cambridge University Botanic Gardens, Cambridgeshire, UK (Comba *et al.* 1999a) found that bumblebee visits to hollyhock *Alcea rosea* and larkspur *Consolida* sp. were more frequent on natural, single-petalled forms than on horticulturally modified, double-petalled varieties. Bee visits to four of the flower types – nasturtium *Tropaeolum majus*, pansy *Viola* x *wittrockiana*, marigold *Tagetes patula* and snapdragon *Antirrhinum majus* – were infrequent despite ample nectar provision from some varieties. There was a tendency for wild bees to prefer natural flower shapes in pansy, marigold and snapdragon, but not in nasturtium.

A trial of 25 native flowering herb species planted in the Cambridge University Botanic Gardens, UK, identified 16 species frequently visited by wild bees (Comba *et al.* 1999b). Ten species (seven of which were frequently visited by wild bees) were shown to provide abundant nectar in the garden environment.

A trial of six native plant species (marsh woundwort *Stachys palustris*, wood betony *S. officinalis*, purple loosestrife *Lythrum salicaria*, common toadflax *Linaria vulgaris*, bird's-foot trefoil *Lotus corniculatus* and meadow clary *Salvia pratensis*) recommended for pollinator-friendly gardens in the Cambridge University Botanic Gardens, UK, found all six were nectar-rich and frequently visited by wild bees (Corbet *et al.* 2001). A double-flowered variant of bird's-foot trefoil tested in the same study produced no nectar and attracted no insects.

A replicated trial in the Phoenix metropolitan area, in the Sonoran Desert of the USA, found that eight gardens planted with dry-loving plants (xeric landscaping) supported a greater diversity of bees than eight gardens planted with non-native plants such as grasses that needed to be irrigated (McIntyre & Hostetler 2001). In September, xeric gardens had approximately 10 bee species, compared to less than five species/garden in ordinary gardens.

Tommasi *et al.* (2004) measured bee abundance and diversity in wild areas, and gardens managed for wildlife or managed traditionally, in Vancouver, British Columbia, Canada (five to eight sites of each type). They found that gardens managed for wildlife under the 'Naturescape' programme, sown with native plant species and infrequently mown grass areas, had significantly more bee individuals than traditionally managed gardens with mown lawns and non-native plants such as petunia *Petunia* sp., tulip *Tulipa* sp., pansy *Viola* sp. and rhododendron *Rhododendron* sp. (approximately 45 wild bee (non-*Apis mellifera*) individuals caught/hour of sampling on average in Naturescape gardens, compared to less than 20 bees/hour in traditional gardens). Naturescape gardens did not have significantly more bee species than traditional gardens.

Wojcik *et al.* (2008) planted a 180 m^2 urban plot at the University of California, Berkeley, USA, with 78 garden plants chosen to provide a consistent floral resource throughout the season, and monitored bee visits the following summer. The plot provided pollen and nectar from spring to early autumn, and attracted 32 bee species, from 17 genera and five families, demonstrating the potential of newly planted urban gardens to provide resources for native bees.

Pawelek *et al.* (2009) added 41 types of 'bee attractive' plants, both native and non-native, to a 4,000 m^2 community garden in San Luis Obispo, California and monitored the abundance and diversity of native bees over three years from 2007 to 2009. Plants were planted in 1 × 1.5 m patches, in 19 of the 29 plots within the garden, and bees were recorded in 3-minute counts on selected patches, regularly between July and October. The number of bee species recorded rose from five in 2007 (less sampling effort than subsequent years) to 21 in 2008 and 31 (including four non-native species) in 2009. The added plants that attracted the greatest number of wild bee species were blanketflower *Gaillardia x grandiflora* and bog sage *Salvia uliginosa* (both non-native, 11 species recorded on each).

Comba L., Corbet S.A., Barron A., Bird A., Collinge S., Miyazaki N. & Powell M. (1999a) Garden flowers: Insect visits and the floral reward of horticulturally-modified variants. *Annals of Botany*, 83, 73–86. www↗

Comba L., Corbet S.A., Hunt L. & Warren B. (1999b) Flowers, nectar and insect visits: eval-
uating British plant species for pollinator friendly gardens. *Annals of Botany*, 83,
369–383. www↗

Corbet S.A., Bee J., Dasmahapatra K., Gale S., Gorringe E., La Ferla B., Moorhouse T., Trevail
A., Van Bergen Y. & Vorontsova M. (2001) Native or exotic? Double or single? Evaluat-
ing plants for pollinator-friendly gardens. *Annals of Botany*, 87, 219–232. www↗

McIntyre N.E. & Hostetler M.E. (2001) Effects of urban land use on pollinator (Hymenoptera: Apoidea) communities in a desert metropolis. *Basic and Applied Ecology*, 2,
209–218. www↗

Pawelek J.C., Frankie G.W., Thorp R.W. & Przybylski M. (2009) Modification of a community
garden to attract native bee pollinators in urban San Luis Obispo, California. *Cities
and the Environment*, 2, article 7. www↗

Tommasi D., Miro A., Higo H.A. & Winston M.L. (2004). Bee diversity and abundance in an
urban setting. *The Canadian Entomologist*, 136, 851–869. www↗

Wojcik V.A., Frankie G.W., Thorp R.W. & Hernandez J.L. (2008) Seasonality in bees and
their floral resource plants at a constructed urban bee habitat in Berkeley, California.
Journal of the Kansas Entomological Society, 81, 15–28. www↗

2.2 Practise 'wildlife gardening'

• A site comparison study in one city in the UK found more species of bumblebee in
domestic city gardens with lower intensity of management, a measure reflecting
the tidiness of the garden and the use of garden pesticides. Solitary bees were not
affected by this measure.

Background

Members of the public are encouraged to manage their gardens for wildlife by planting appropriate plants (see Chapter 1.1 *Plant parks and gardens
with appropriate flowers*), leaving areas of the garden unmanaged, keeping
ponds and compost heaps, providing nest boxes and food for wildlife
and reducing inputs of herbicides and pesticides. This set of techniques is
generally known as 'wildlife gardening' or 'habitat gardening'. Aspects
of it other than planting forage plants, such as reducing pesticide use and
leaving areas unkempt, are also interventions for bee conservation.

A site comparison study of 61 domestic gardens in the city of Sheffield, UK, recorded the abundance and species richness of invertebrates, including bees, along with aspects of garden management, based on questionnaires issued to householders (Smith *et al.* 2006a, b). Bees were sampled in flight using a 'Malaise trap' set for two weeks between June and September in each garden. 'Management intensity' was calculated from scores for weeding, pruning, watering, dead-heading flowers, collecting autumn leaves, and the use of fertilizers, herbicides, pesticides. An index of 'wildlife management' was based on whether householders fed birds, provided bird nest boxes or used other (unspecified) methods to attract wildlife. Results showed that the number of bumblebee species (but not their abundance) was related to management intensity, with more species in gardens managed less intensively. Solitary bee species richness was related to the number of plant species in the garden (both native and alien), but not directly to the interventions relevant to wildlife gardening. The abundance of solitary bees was related to the number of native plant species, but was lower in gardens with a higher index of 'wildlife gardening' (mostly focussed on encouraging birds).

Smith R.M., Warren P.H., Thompson K. & Gaston K.J. (2006a) Urban domestic gardens (VI): environmental correlates of invertebrate species richness. *Biodiversity and Conservation*, 15, 2415–2438. www↗

Smith R.M., Gaston K.J., Warren P.H. & Thompson K. (2006b) Urban domestic gardens (VIII): environmental correlates of invertebrate abundance. *Biodiversity and Conservation*, 15, 2515–2545. www↗

2.3 Protect brownfield sites

- We have captured no evidence for the effects of interventions to protect brownfield sites from insensitive re-development.

Background

'Brownfield sites' are ex-industrial or previously developed sites that have been abandoned. They can support a high diversity of insects, including bees and in some cases threatened or declining species. In the

UK, these sites are a target for development and they have become a focus of attention for insect conservationists.

2.4 Conserve old buildings or structures as nesting sites for bees

- We have captured for evidence on the effects of conserving old buildings and structures suitable for nesting wild bees.

Background

Anecdotally, old buildings can represent valuable nesting sites for wild bees, particularly cavity-nesting species. Efforts to conserve bees can involve retaining or delaying renovation of such buildings.

3. Threat: land use change due to agriculture

Protect existing natural or semi-natural habitat to prevent conversion to agriculture

We have captured no evidence for the effects of protecting areas of natural or semi-natural habitat on bee populations or communities.

Increase the proportion of natural or semi-natural habitat in the farmed landscape

We know of no evidence demonstrating the effects of restoring natural or semi-natural habitat on bee diversity or abundance in surrounding farmland.

Provide set-aside areas in farmland

Two replicated trials in Germany showed that species richness of bees nesting or foraging (one study for each) is higher on set-aside left for two years or more, relative to other management regimes or, in the nesting study, arable fields.

Restore species-rich grassland vegetation

One replicated controlled trial in Scotland showed that species-rich grassland managed under agri-environment schemes attracted more nest-searching queen bumblebees but fewer foraging queens in the spring than unmanaged grassland. Three small trials in the UK or Germany found that restored species-rich grasslands had similar flower-visiting insect communities to paired ancient species-rich grasslands.

Restore heathland

Two replicated UK trials indicated that long-term restoration of dry lowland heath can restore a bee community similar to that on ancient heaths, after 10–14 years. We found no evidence on interventions to conserve bees on upland heaths.

Connect areas of natural or semi-natural habitat

We have captured no evidence of the effects on wild bee communities of connecting patches of natural or semi-natural habitat.

Reduce tillage

Two replicated trials on squash farms in the USA had contrasting results. One showed no difference in the abundance of bees between tilled and untilled farms, the other found three times more squash bees *Peponapis pruinosa* on no-till than on conventional farms.

Increase areas of rough grassland for bumblebee nesting

One replicated controlled trial on lowland farms in Scotland showed that grassy field margins attracted nest-searching queen bumblebees in spring at higher densities than cropped field margins, managed or unmanaged grasslands or hedgerows.

Create patches of bare ground for ground-nesting bees

One replicated controlled trial in Germany and four small trials elsewhere in Europe or North America (three replicated, one not) have shown that artificially exposed areas of bare soil can be successfully colonised by ground-nesting solitary bees and wasps in the first or second year.

Provide grass strips at field margins

Three replicated, controlled trials in the UK have monitored wild bees on uncropped grassy field margins. Evidence of the effects on bees is mixed. One trial showed that grassy field margins enhanced the abundance, but not diversity, of wild bees at the field boundary. One showed that grassy field margins enhanced the abundance and diversity of bumblebees within the margin. A third, smaller-scale trial showed neither abundance nor diversity of bumblebees was higher on sown grassy margins than on cropped margins.

Manage hedges to benefit bees

One replicated controlled trial shows that hedges managed under the Scottish Rural Stewardship scheme do not attract more nest-searching or foraging queen bumblebees in spring than conventionally managed hedgerows.

Increase the use of clover leys on farmland

We have captured no evidence that increasing the use of clover leys can enhance wild bee populations.

Plant dedicated floral resources on farmland

Fourteen trials in Europe and North America have recorded substantial numbers of wild bees foraging on sown flowering plants in farmland. Four replicated trials monitored the wider response of bee populations by measuring reproductive success, numbers of nesting bees or numbers foraging in the surrounding landscape. One, in Canada, found enhanced reproductive success of blue orchard mason bees *Osmia lignaria*.

Sow uncropped arable field margins with an agricultural 'nectar and pollen' mix

Five replicated trials in Europe have documented bumblebees foraging on field margins sown with an agricultural nectar and pollen seed mix. Four showed that planted legumes attract significantly more bumble-bees than naturally regenerated, grassy or cropped field margins. Three showed that they attract more bumblebees than a perennial wildflower mix, at least in the first year.

Sow uncropped arable field margins with a native wild flower seed mix

Five replicated trials in the UK have shown that uncropped field margins sown with wild flowers support a higher abundance (and in three trials species richness) of foraging bumblebees than cropped field edges (all five trials), grassy margins (four trials) or naturally regenerated un-cropped margins (three trials).

Leave arable field margins uncropped with natural regeneration

Four replicated trials in the UK have found more bumblebees and/or bee species on uncropped field margins than on cropped margins. One small replicated UK trial found neither abundance nor diversity of bumblebees was higher on naturally regenerated than on cropped margins.

Increase the diversity of nectar and pollen plants (including crop plants) in the landscape

One large replicated controlled UK trial showed that the abundance of long-tongued bumblebees on field margins was positively correlated with the number of 'pollen and nectar' agri-environment agreements in a 10 km grid square.

Reduce the intensity of farmland meadow management

Four replicated trials in Europe have compared farmland meadows man-aged extensively with conventionally farmed meadows or silage fields. Two found enhanced numbers and diversity of wild bees on meadows with a delayed first cut and little agrochemical use. Two found no differ-ence in bee diversity or abundance.

Reduce grazing intensity on pastures

One replicated trial in Germany has shown that reducing the intensity of summer cattle grazing can increase the abundance, but not the species richness of cavity-nesting bees and wasps.

For all evidence relating to the use of **nest boxes**, see Chapter 10 *Provide artificial nest sites for bees*.

For the effects of converting to **organic farming**, and studies that moni-tored the effects of several different **agri-environment schemes** at once, see Chapter 4 *Threat: pollution - agricultural and forestry effluents*.

Background

Land use changes due to agriculture, particularly intensification of agriculture, natural habitat fragmentation and the abandonment of traditional practices, are significant drivers of declines in pollinator diversity (Kuldna *et al.* 2009).

Threats from wood and pulp plantations are also included in this category.

How we treat European agri-environment schemes

In Europe, agri-environment schemes represent a crucial instrument for intervening to support wildlife in the farmed environment. They compensate farmers financially for changing agricultural practice to be more favourable to biodiversity and landscape. The schemes are an integral part of the European Common Agricultural Policy and Member States devise their own agri-environment prescriptions to suit their agricultural economies and environmental contexts.

Since agri-environment schemes represent many different specific interventions relevant to bee conservation, they fall into different sections in this synopsis, appearing both in this chapter and in the subsequent chapter on agricultural effluents. The placement of interventions is pragmatic. We have placed agri-environment schemes that have a strong component of reduced chemical use, such as organic farming, in the chapter on pollution. We also place studies that look at a range of different agri-environment schemes in this chapter, because the different prescriptions have reduced chemical use in common. Clearly many of these schemes alter aspects of landscape and habitat as well, so their effects cannot be entirely attributed to the change in chemical use.

Overall, there is a substantial amount of complex evidence relating to the effects of agri-evironment schemes. We recommend a systematic review that brings together all the interventions included as agri-environment schemes in Europe and assesses the evidence relating to their effect on bees, or pollinators more widely.

The importance of measuring population effects rather than forager numbers

Many trials of planting floral resources in farmland for bees (for example, in field margins, or dedicated plots) measure only the numbers of foraging bees visiting the planted flowers. It is important to note that, since some bees have foraging ranges up to 2 km or more, increased numbers of bees or bee species at flowers may just reflect a redistribution of individuals in the landscape rather than any population-level effects.

It is even conceivable that concentrating bees on field margins has an adverse effect on bees, drawing them in from the surrounding landscape to an area where they are at greater risk of exposure to pesticides (we know of no evidence for this).

For these reasons, we particularly highlight the small number of studies that monitor other effects of planting floral resources, such as bee numbers in the surrounding landscape, or numbers of nesting bees. We do, however, include evidence on the use of planted floral resources by foraging bees, even in cases where there has been no control plot. We consider this to be the first stage of evidence that the intervention is effective.

Bee populations, like those of most insects, vary greatly from year to year. Measuring genuine changes in bee populations requires monitoring over at least five years, at scales of several km^2.

Techniques for creating bee habitat

We have not included studies unless they have directly monitored the effects of interventions on bees. This means we have excluded evidence concerning management techniques to restore and maintain bee habitat, such as experiments on how to restore species-rich grassland (reviewed for the UK by Walker *et al.* 2004), or establish clover forage (for example Allcorn *et al.* 2006). This information will be included in forthcoming synopses centred on habitat management.

Allcorn R.I. Akers P. & Lyons G. (2006) Introducing red clover *Trifolium pratense* to former arable fields to provide a foraging resource for bumblebees *Bombus* spp. at Dungeness RSPB reserve, Kent, England. *Conservation Evidence* 3, 88–91. www↗

Kuldna P., Peterson K., Poltimae H. & Luig J. (2009) An application of DPSIR framework to identify issues of pollinator loss. *Ecological Economics*, 69, 32–42.

Walker K.J., Stevens P.A., Stevens D.P., Mountford J.O., Manchester S.J. & Pywell R.F. (2004) The restoration and re-creation of species-rich lowland grassland on land formerly managed for intensive agriculture in the UK. *Biological Conservation*, 119, 1–18.

3.1 Protect existing natural or semi-natural habitat to prevent conversion to agriculture

• We have captured no evidence for the effects of protecting areas of natural or semi-natural habitat on bee populations or communities.

Background

Protecting areas of existing habitat from conversion to agriculture is one of the most important conservation measures, particularly in the tropics. Whilst there is evidence that establishing protected areas reduces the rate of habitat degradation (see for example Bruner *et al.* 2001, Gaston *et al.* 2008), we do not know of any specific evidence demonstrating that protected areas are effective at enhancing or protecting wild bee populations.

Several studies show that bee abundance and/or diversity on farms are higher when areas of natural or semi-natural habitat are closer, or the proportion of natural or semi-natural habitat in the surrounding landscape is higher (reviewed by Ricketts *et al.* 2008, for example). But these studies have not monitored bee numbers in response to specific interventions to protect habitat.

Bruner A.G., Gullison R.E., Rice R.E. & da Fonseca G.A.B. (2001) Effectiveness of parks in protecting tropical biodiversity. *Science*, 291, 125–128.

Gaston K.J., Jackson S.E., Cantu-Salazar L. & Cruz-Pinon G. (2008) The ecological performance of protected areas. *Annual Review of Ecology Evolution and Systematics*, 39, 93–113.

Ricketts T. H., Regetz J., Steffan-Dewenter I., Cunningham S.A., Kremen C., Bogdanski A., Gemmill-Herren B., Greenleaf S.S., Klein A.M, Mayfield M.M., Morandin L.A., Ochieng A., Potts S.G. & Viana B.F. (2008) Landscape effects on crop pollination services: are there general patterns? *Ecology Letters*, 11, 499–515.

3.2 Increase the proportion of natural or semi-natural habitat in the farmed landscape

- We know of no evidence demonstrating the effects of restoring natural or semi-natural habitat on bee diversity or abundance in neighbouring farms.

Several studies show that bee abundance and/or diversity on farms are higher when areas of natural or semi-natural habitat (including forest) are closer or the proportion of natural or semi-natural habitat in the surrounding landscape is higher (reviewed by Ricketts *et al.* 2008, for example). We know of no evidence demonstrating that the restoration of natural or semi-natural habitats affects bee diversity or abundance in neighbouring farms, although at least one such study is currently underway.

Ricketts T. H., Regetz J., Steffan-Dewenter I., Cunningham S.A., Kremen C., Bogdanski A., Gemmill-Herren B., Greenleaf S.S., Klein A.M, Mayfield M.M., Morandin L.A., Ochieng A., Potts S.G. & Viana B.F. (2008) Landscape effects on crop pollination services: are there general patterns? *Ecology Letters*, 11, 499–515.

3.3 Provide set-aside areas in farmland

- Two replicated trials showed that species richness of bees nesting (one study) or foraging (one study) is higher on set-aside that is annually mown and left to naturally regenerate for two years or more, relative to other set-aside management regimes or, in the nesting study, to arable crop fields.

A replicated, controlled trial with four replicates of each treatment (Gathmann *et al.* 1994) compared cavity-nesting bees and wasps nesting on set-aside arable land managed in six different ways with crop fields and old meadows in Kraichgau, southwest Germany. The study used reed *Phragmites australis* stem nest boxes (described in 'Provide artificial nest sites for solitary bees'), and recorded nesting only, not foraging activity. Set-aside fields were either sown in the year of study, with a grass-clover mix or phacelia *Phacelia tanacetifolia* (also known as scorpion weed, lacy phacelia or tansy phacelia) or were in their first or second year of natural regeneration, with or without mowing. Overall, naturally regenerated fields had significantly more nests, and more nesting species than fields sown with fallow or arable crops.

Of the six set-aside treatments, the most species were found on two-year-old set-aside, mown in late June or early July, with a total of eight nesting bee species. This compares with four bee species found on 1-year-old unmown set-aside, and none on set-aside sown with phacelia. Twelve bee species were found on old meadows (>30 years old, with old fruit trees). Amongst 2-year-old, naturally regenerated set-aside fields, mown fields had more than twice as many species (bees and wasps) as unmown fields (average 4.8 species/field versus 1.8).

A second replicated trial in the same region (Steffan-Dewenter & Tscharntke 2001) examined the abundance and species richness of foraging bees, both solitary and social, on annually mown set-aside fields of different ages and management. The number of bee species increased with the age of set-aside fields, from 15 species on 1-year-old fields to 25 species on 5-year-old fields. Two-year-old set-aside fields had the most bee species – 29 on average, compared to 32 species for old meadows, including an average of around five oligolectic species (specialising on pollen from a small group of plant species). One-year-old set-aside fields sown with phacelia had an average of 13 bee species, mainly common, generalised species of bumblebee *Bombus* and *Lasioglossum*.

Gathmann A., Greiler H-J. & Tscharntke T. (1994) Trap-nesting bees and wasps colonizing set-aside fields: succession and body size, management by cutting and sowing. *Oecologia*, 98, 8–14. www↗

Steffan-Dewenter I. & Tscharntke T. (2001) Succession of bee communities on fallows. *Ecography*, 24, 83–93. www↗

3.4 Restore species-rich grassland vegetation

See also Chapter 5.1 *Restore species-rich grassland on road verges*

- One replicated controlled trial in Scotland showed that species-rich grassland managed under agri-environment schemes attracted more nest-searching queen bumblebees but fewer foraging queens in the spring than unmanaged grassland.

- Three small trials, two in the UK and one in Germany, found that restored species-rich grasslands had similar flower-visiting insect communities (dominated by bees and/or flies) to paired ancient species-rich grasslands.

A study in eastern England of the pollinator community on a species-rich grassland restoration experiment compared to native grassland of the same plant community found a greater diversity of pollinating insects on the restored hay meadow site than on the ancient meadow (Dicks 2002). Six common species of bumblebee were recorded at both sites, and the most abundant insect visitor was a bumblebee on both meadows: white-tailed bumblebees *Bombus terrestris/ lucorum* at the restored site, red-tailed bumblebees *B. lapidarius* at the ancient meadow site. Seven and five species of solitary bee were recorded at restored and ancient sites respectively.

A comparison of two restored hay meadows with two ancient hay meadows in the Bristol area, UK (Forup & Memmott 2005) found no consistent differences in the abundance or diversity of pollinating insects (dominated by bees and flies) between ancient and restored sites, and considered the pollinator community to be effectively restored.

A replicated, controlled trial of the species-rich grassland management or restoration option under the Rural Stewardship agri-environment scheme in Scotland (Rural Stewardship Scheme, RSS) found that RSS species-rich grassland attracted more nest-searching queen bumblebees *Bombus* spp. but fewer foraging queens than areas of naturally regenerated, largely unmanaged grasslands (Lye *et al.* 2009). Five RSS farms were paired with five conventional farms. Across all farms, unmanaged grassland on conventional farms attracted the highest abundance of foraging queen bumblebees (over 4 queens/100 m transect, compared to less than 3 foraging queens/100 m transect on species-rich grassland), also in comparison with hedgerow and field margin transects. Unmanaged grassland transects had more nectar and pollen-providing flowers than species-rich grassland in April and May, when queen bumblebees are on the wing.

A comparison of two restored sandy grassland and riverine sand dune complexes with the target semi-natural grassland communities near the River Hase, Lower Saxony, Germany found no significant difference in the number of bee species between target and restored sites in any study year, two to five years after restoration (Exeler *et al.* 2009). Bees were more abundant at semi-natural sand dunes than at restored sand dune sites, but this was not true for the semi-natural sandy grassland sites, characterised by maiden pink *Dianthus deltoides* and thrift *Armeria elongata*.

Dicks L.V. (2002) The structure and functioning of flower-visiting insect communities on hay meadows. PhD thesis, University of Cambridge. www↗

Exeler N., Kratochwil A. & Hochkirch A. (2009) Restoration of riverine inland sand dune complexes: implications for the conservation of wild bees. *Journal of Applied Ecology*, 46, 1097–1105. www↗

Forup M.L. & Memmott J. (2005) The restoration of plant-pollinator interactions in hay meadows. *Restoration Ecology*, 13, 265–274. www↗

Lye G., Park K., Osborne J., Holland J. & Goulson D. (2009) Assessing the value of Rural Stewardship schemes for providing forage resources and nesting habitat for bumblebee queens (Hymenoptera: Apidae). *Biological Conservation*, 142, 2023–2032. www↗

3.5 Restore heathland

• One small trial of early-stage lowland heath restoration activity did not have an adverse effect on bumblebee diversity at one site in southeast England. Two replicated trials in Dorset indicated that long-term restoration of dry lowland heath can restore a bee community similar to that on ancient heaths. One of these studies showed that the community of conopid flies parasitizing bumblebees remained impoverished 15 years after heathland restoration began. We found no evidence on interventions to conserve bees on upland heath or moorland.

Selective tree felling and removal of humus and nutrient-rich soil by scraping in a 1 ha area at Norton Heath Common, southeast England increased the range of common bumblebee species recorded within the scraped area from one in the first year to four in the second year (Gardiner & Vaughan 2008).

Forup *et al.* (2008) compared four ancient dry lowland heaths in Dorset with four paired heathland sites first restored from pine *Pinus* sp. plantation 11 to 14 years previously. There were no consistent differences between the communities of insect pollinators, including bees, at ancient and restored sites. There was no clear evidence that bees or other pollinators colonised restored heaths from the adjacent or nearby paired ancient heaths, implying that from a bee perspective, there is no need to site heathland restoration projects very close to ancient sites.

Henson *et al.* (2009) sampled bumblebees visiting flowers on six ancient and six restored patches of heathland on the Isle of Purbeck, Dorset, UK. The restored sites had been restored from pine plantation around 10 years previously. The species richness and abundance of bumblebees were similar on ancient and restored sites, as were those of bumblebee protozoan parasites, external and tracheal mites. But conopid flies, a type of internal bumblebee parasitoid, were significantly less abundant on restored sites than ancient sites.

Forup M.L., Henson K.S.E., Craze P.G. & Memmott J. (2008) The restoration of ecological interactions: plant-pollinator networks on ancient and restored heathlands. *Journal of Applied Ecology*, 45, 742–752. www↗

Gardiner T. & Vaughan A. (2008) Responses of ground flora and insect assemblages to tree felling and soil scraping as an initial step to heathland restoration at Norton Heath Common, Essex, England. *Conservation Evidence*, 5, 95–100. www↗

Henson K.S.E., Craze P.G. & Memmott J. (2009) The restoration of parasites, parasitoids, and pathogens to heathland communities. *Ecology*, 90, 1840–1851. www↗

3.6 Connect areas of natural or semi-natural habitat

• We have captured no evidence of the effects on wild bee communities of connecting patches of natural or semi-natural habitat.

There is evidence that pollination services can be enhanced when patches of semi-natural habitat are connected (for example, Townsend & Levy 2005). However, we have not found evidence of the effects of connecting areas of natural or semi-natural habitat together on wild bee populations or communities.

Townsend P.A. & Levey D.J. (2005) An experimental test of whether habitat corridors affect pollen transfer. *Ecology*, 86, 466–475.

3.7 Reduce tillage

- Evidence on whether reduced tillage or no tillage benefits ground-nesting bees is mixed. Two replicated trials on squash *Cucurbita* spp. farms in the USA had contrasting results. One showed no difference in the abundance of bees between tilled and untilled farms, the other found three times more squash bees *Peponapis pruinosa* on no-till farms than on conventional farms.

Background

Tillage might be expected to impact on the immature stages of ground-nesting bees, by breaking up nests and damaging larvae, so reducing tillage depth or practising no-till farming could benefit these bees.

The squash bee *Peponapis pruinosa* is a test case for this because it is known to nest near its host plant, so it often nests within crop fields rather than in field margins.

Two replicated trials have compared the effects of tillage on the abundance of squash bee and other bees visiting squash *Cucurbita* spp. flowers in the United States. Both studies used 20 or more farms, in the same area (Virginia or Maryland, USA). Shuler *et al.* (2005) found that there were three times more squash bees on no-till farms as on tilled farms, although there was no difference in the numbers of bumblebees *Bombus* spp. or honey bees *Apis mellifera*. By contrast, Julier & Roulston (2009) found no difference in the numbers of squash bees or other bees between farms that had tilled after the previous year's pumpkin crop and those that had not. Julier & Roulston's study only included farms growing pumpkins, which are relatively late flowering compared to other cultivated squash plants. Early emerging squash bees may have been missed by this study because they had to travel elsewhere to forage and nest.

Shuler R.E., RoulstonT.H. & Farris G.E. (2005). Farming practices influence wild pollinator populations on squash and pumpkin. *Journal of Economic Entomology*, 98, 790–795. www↗

Julier H.E. & Roulston T.H. (2009). Wild bee abundance and pollination service in cultivated pumpkins: farm management, nesting behaviour and landscape effects. *Journal of Economic Entomology*, 102, 563–573. www↗

3.8 Increase areas of rough grassland for bumblebee nesting

- One replicated controlled trial on lowland farms in Scotland showed that grassy field margins attracted nest-searching queen bumblebees in spring at higher densities than cropped field margins, managed or unmanaged grasslands or hedgerows.

A replicated, controlled trial of the Rural Stewardship agri-environment scheme on 10 farms in Scotland found that 1.5 to 6 m wide grassy field margins attracted nest-searching queen bumblebees at higher densities than managed or unmanaged grasslands or hedgerows (Lye *et al.* 2009). On five farms with the agri-environment scheme, researchers counted an average of around nine nest-searching queens/100 m on grassy field margins, compared to around seven nest-searching queens/100 m in species-rich grassland transects, five for conventional arable field margins, and four on unmanaged (abandoned) grassland transects. The study did not record the numbers of established nests later in the year.

Lye G., Park K., Osborne J., Holland J. & Goulson D. (2009) Assessing the value of Rural Stewardship schemes for providing foraging resources and nesting habitat for bumblebee queens (Hymenoptera: Apidae). *Biological Conservation*, 142, 2023–2032. www↗

3.9 Create patches of bare ground for ground-nesting bees

- One replicated controlled trial in Germany and four small trials (three replicated, one not) have shown that artificially exposed areas of bare soil can be successfully colonised by ground-nesting solitary bees and wasps in the first or second year. We have captured no evidence for the effect of creating areas of bare ground on bee populations or communities on a larger scale.

Three scrapes with vegetation removed at Headley Heath, Surrey, UK were being used by ground-nesting bees two or three years after they were created (Edwards 1996). The average densities of burrows attributed to ground-nesting bees and wasps were 2.3, 1.2 and 2.3 burrows/m² for small (500 m²), medium (2,500 m²) and large (5,000 m²) scrapes respectively

Nest density of ground-nesting bees and wasps was increased by removing plant cover, or digging and raising soil, in trial plots at five sandy grassland sites in Baden-Württemberg, southwest Germany in 1992, relative to five control sites (Wesserling & Tscharntke 1997). Digging and raising soil was more effective at increasing nest density. Raking was not very effective, because it generated a dense plant cover.

A study of artificially made scrapes on three lowland heaths in West Sussex, UK (Edwards 1998) found between two and eight solitary bee species using the scrapes one to four years after they were created, with up to five of the species actively nesting.

Severns (2004) created 1 m^2 plots of mostly bare ground whilst planting seeds of the endangered legume Kincaid's lupine *Lupinus sulphureus* spp. *kincaidii*, in an upland prairie restoration project in Oregon, USA. The bare ground was colonised by an increasing number of nesting solitary bees, mostly of the common species *Lasioglossum anhypops*, over the following three years. Three years later, there were 320 nests over 30 plots.

Four shallow bays (3 × 5 m), with a rear vertical face (30 cm), were dug to attract ground-nesting bees and wasps at Shotover Hill, a heath degraded due to a long-term lack of grazing in Oxfordshire, southern UK. All four bays were colonised in the first year with 80 solitary bee and wasp species recorded in the following three years (Gregory & Wright 2005).

Edwards M. (1996) Entomological Survey and Monitoring, Headley Heath, 1995–1996. Unpublished report commissioned by The National Trust. www↗

Edwards M. (1998) Monitoring of bare ground for use by heathland insects. Unpublished report to the West Sussex Heathlands Project. www↗

Gregory S. & Wright I. (2005) Creation of patches of bare ground to enhance the habitat of ground-nesting bees and wasps at Shotover Hill, Oxfordshire, England. *Conservation Evidence*, 2, 139–141. www↗

Severns P. (2004) Creating bare ground increases presence of native pollinators in Kincaid's lupine seeding plots (Oregon). *Ecological Restoration*, 22, 234–235. www↗

Wesserling J. & Tscharntke T . (1995) Habitat selection of bees and digger wasps – experimental management of plots. *Mitteilungen der Deutschen Gesellschaft für Allgemeine und Angewandte Entomologie*, 9, 697–701. www↗

3.10 Provide grass strips at field margins

See also Chapter 3.8 *Increase areas of rough grassland for nesting bumblebees*, for a study of the use of grassy margins by nest-searching queen bumblebees and Chapter 4.1 *Introduce agri-environment schemes that reduce spraying*.

- Three replicated controlled trials in the UK have monitored wild bees on uncropped grassy field margins. Evidence of the effects on bees is mixed. One trial showed that 6 m wide grassy field margins enhanced the abundance, but not diversity, of wild bees at the field boundary. One trial showed that 6 m wide grassy field margins enhanced the abundance and diversity of bumblebees within the margin. A third, smaller scale trial showed neither abundance nor diversity of bumblebees was higher on sown grassy margins than on cropped margins.

A small replicated, controlled trial of field margin management options on two farms in North Yorkshire, UK in one summer (Meek *et al.* 2002) did not find significantly more bumblebees on margins sown with tussocky grass than on naturally-regenerated margins or cropped margins. There were four replicates of each treatment.

A replicated, controlled trial of the 6 m wide grassy field margin agri-environment scheme option at 21 sites in England found no difference in the diversity of wild bees (sampled in the field boundary by walked transect and sweep netting) between paired control fields and fields with sown grassy margins (Kleijn *et al.* 2006).

The same study, reported elsewhere (Marshall *et al.* 2006), showed a significantly greater abundance of bees in boundaries of fields with sown grassy margins; 40% of the bees recorded were of one species, the red-tailed bumblebee *Bombus lapidarius*.

A replicated, controlled trial of the 6 m wide sown grassy field margin agri-environment option at 32 sites across England (Pywell *et al.* 2006) found that grassy margins had more species, and a higher abundance of foraging bumblebees, than conventionally cultivated and cropped field margins (on average 6–8 bees of 1.3–1.4 species per transect on grassy margins, compared to 0.2 bees of 0.1 species/transect for cropped margins). Older grassy margins, sown more than three years previously, did not attract more foraging bumblebees than those sown in the previous two years.

Kleijn D., Baquero R.A., Clough Y., Diaz M., De Esteban J., Fernandez F., Gabriel D., Herzog F., Holzschuh A., Johl R., Knop E., Kruess A., Marshall E.J.P., Steffan-Dewenter I., Tscharntke T., Verhulst J., West T.M. & Yela J.L. (2006) Mixed biodiversity benefits of agri-environment schemes in five European countries. *Ecology Letters*, 9, 243–254. www↗

Marshall E.J.P., West T.M. & Kleijn D. (2006) Impacts of an agri-environment field margin prescription on the flora and fauna of arable farmland in different landscapes. *Agriculture, Ecosystems and Environment*, 113, 36–44. www↗

Meek B., Loxton D., Sparks T., Pywell R., Pickett H. & Nowakowski M. (2002) The effect of arable field margin composition on invertebrate biodiversity. *Biological Conservation*, 106, 259–271. www↗

Pywell R.F., Warman E.A., Hulmes L., Hulmes S., Nuttall P., Sparks T.H., Critchley C.N.R. & Sherwood A. (2006) Effectiveness of new agri-environment schemes in providing foraging resources for bumblebees in intensively farmed landscapes. *Biological Conservation*, 129, 192–206. www↗

3.11 Manage hedges to benefit bees

- One replicated controlled trial showed that hedges managed under the Scottish Rural Stewardship scheme do not attract more nest-searching or foraging queen bumblebees in spring than conventionally managed hedgerows.

A replicated, controlled trial of the Rural Stewardship agri-environment scheme on five farms in Scotland found that hedgerows dominated by hawthorn *Crataegus monogyna* or blackthorn *Prunus spinosa* were less attractive than field margins or grasslands to nest-searching queen bumblebees *Bombus* spp. in April and May (Lye *et al.* 2009). There was no significant difference in numbers of foraging or nesting queens between hedgerows managed under the agri-environment scheme (winter cut every three years, gaps filled, vegetation below unmown and unsprayed) and conventionally managed hedgerows. The study took place before the woody species comprising the hedgerow came into flower.

Lye G., Park K., Osborne J., Holland J. & Goulson D. (2009) Assessing the value of Rural Stewardship schemes for providing foraging resources and nesting habitat for bumblebee queens (Hymenoptera: Apidae) *Biological Conservation*, 142, 2023–2032. www↗

3.12 Increase the use of clover leys on farmland

- We have captured no evidence that increasing the use of clover leys can enhance wild bee populations. One replicated trial in Germany showed that fields planted with a white clover grass mixture do not attract solitary bees to nest preferentially on site. A trial in Switzerland showed that if white clover is mowed during flowering, injuries and mortality of bees can be reduced by avoiding the use of a processor attached to the mower.

As part of a larger study with 10 field types, Gathmann *et al.* (1994) placed bundles of reed stems *Phragmites australis* for cavity-nesting bees (and wasps) in four set-aside fields newly sown with a clover-grass mix. The mix was mostly white clover *Trifolium repens*, perennial rye grass *Lolium perenne* and alfalfa *Medicago sativa*. Four species of bee made nests in the reed stems in these fields, including one endangered species *Megachile alpicola*. However, in the same study, three of those four species also nested in reed stems placed in barley *Hordeum vulgare* fields. By contrast, 12 bee species nested in reed stems placed in 2-year-old set-aside fields mown in late June, and 16 species nested in reed stems in old meadows.

Fluri & Frick (2002) rotary mowed white clover crops during flowering with and without a mechanical processor, and monitored the death and injury to actively foraging honey bees *Apis mellifera*, on two 0.33 ha trial plots in Switzerland. During mowing with a rotary mower and processor (which crushes mowings to accelerate drying), 53–62% of the number of foragers recorded before mowing were found injured, dead or otherwise stuck in the mown grass after mowing. When mowing was conducted without a processor, the average number of bees left dead or unable to fly was reduced from an average of 1.4 bees/m^2 (with processor) to 0.2 bees/m^2 and many bees were observed foraging or flying away after passing through the mower. The effects of mowing with a processor (but not without) were also tested on a similar-sized plot of *Phacelia tanacetifolia*, on which bumblebees were recorded as well as honey bees. On average, 0.2 foraging bumblebees/m^2 were recorded before mowing, and 'practically' no bumblebees were found in the mown grass.

Gathmann A., Greiler H-J. & Tscharntke T. (1994) Trap-nesting bees and wasps colonizing set-aside fields: succession and body size, management by cutting and sowing. *Oecologia*, 98, 8–14. www↗

Fluri P. & Frick R. (2002) Honey bee losses during mowing of flowering fields. *Bee World*, 83, 109–118. www↗

3.13 Plant dedicated floral resources on farmland

See also Chapter 3.14 *Sow uncropped arable field margins with an agricultural nectar and pollen mix.*

- Fourteen trials in Europe and North America have recorded substantial numbers of wild bees foraging on perennial or annual sown flowering plants in the agricultural environment.

- Ten trials (eight replicated) have monitored bees foraging on patches sown with a high proportion of phacelia *Phacelia tanacetifolia* on farmland and all but one found substantial numbers of foraging wild (non-*Apis*) bees, particularly bumble-bees *Bombus* spp. Six of these trials recorded the number of foraging bee species, which ranged from eight to 35. One replicated trial shows that phacelia is not very attractive to wild bees in Greece.

- One replicated controlled trial in the UK showed that planted perennial leguminous herbs, including clovers, were more attractive to bumblebees in landscapes with a greater proportion of arable farming.

- Four replicated trials have quantified the wider response of wild bee populations to planted flower patches by measuring reproductive success, numbers of nesting bees or numbers foraging in the surrounding landscape. One trial showed that planted patches of bigleaf lupine *Lupinus polyphyllus* in commercial apple orchards in Novia Scotia, Canada, significantly enhanced the reproductive success of blue orchard mason bees *Osmia lignaria*. One trial in the Netherlands showed that bee numbers and species richness are not higher in farmland 50–1,500 m away from planted flower patches. Two trials in Germany found no or relatively few species of solitary bee nesting on set-aside fields sown with phacelia or clover respectively.

Ten of the studies described below involved planting phacelia *Phacelia tanacetifolia*, a native to California that is often cultivated in Europe as a green manure. Flowering phacelia is very attractive to the largely domesticated honey bee *Apis mellifera*. Here we document evidence of its use by other bee species (wild bees).

Williams & Christian (1991) planted three 9 m² plots of phacelia at Roth-amsted Research experimental farm, Hertfordshire, UK. Seven species of bum-blebee, including the long-tongued common carder bee *Bombus pascuorum*, and one cuckoo bumblebee *B.* [Psithyrus] *vestalis* foraged on the phacelia. Of observed worker bumblebee visits, 97% were for nectar, not pollen.

Patten *et al.* (1993) planted four 1.2 × 1.8 m plots of each of 17 flowering species next to commercial cranberry *Vaccinium macrocarpon* bogs in Washington State, USA. Five plant species attracted more than 30 bees/plot/ count on average: catmint *Nepeta mussini*, borage *Borago officinalis*, phacelia *Phacelia tanacetifolia*, anise hyssop *Agastache foeniculum* and Korean mint *A. rugosa*. Short-tongued bumblebee species *Bombus mixtus, B. occidentalis* and *B. sitkensis* (cranberry pollinators) strongly preferred three plant species: bird's-foot trefoil *Lotus corniculatus*, Korean mint and anise hyssop (averages of 17, 23 and 19 bees/plot/count respectively) but did not visit borage or phacelia much (averages of 1 and 5 short-tongued bees/plot/count, respectively). Two long-tongued species, the Californian bumblebee *B. californicus* and *B. caliginosus* visited borage and phacelia in large numbers (>70 bees/plot/count).

As part of a larger study with 10 field types, Gathmann *et al.* (1994) placed bundles of reed *Phragmites australis* stems for cavity-nesting bees (and wasps) in four set-aside fields newly sown with phacelia in Kraichgau, southwest Germany. The phacelia fields attracted many honey bees *Apis mellifera* (foraging bees not quantified), but no cavity-nesting solitary bees made nests in reed stems in these fields. By comparison, in the same study, 12 bee species nested in reed stems placed in 2-year-old naturally regenerated set-aside fields mown in late June.

Engels *et al.* (1994) planted three strips of the commercially available 'Tübingen nectar and pollen mixture' (40% phacelia, 25% buckwheat *Fagopyron esculentum*) at the edge of an arable field in Baden-Württemberg, Germany, over two years. Two strips were sown only in the first year, one strip was sown in both years. They recorded 58 species of wild bee either nesting in grooved board wooden nest boxes or foraging on the plots, including 11 species of true bumblebee *Bombus* spp. and five species of cuckoo bumblebee *Bombus* [Psithyrus] spp. Thirty-five bee species foraged on flowers from the Tübingen mixture.

Carreck & Williams (1997) planted two or three plots of two commercial nectar and pollen mixtures – Tübingen Mixture (40% phacelia) and Ascot Linde Mixture (25% phacelia) on farmland in Hertfordshire, UK. Across two years, the plots attracted 14 species of bee, including all six common UK bumblebee species and three cuckoo bumblebee species *Bombus* [Psithyrus]. A small number of solitary bees of three species (no more than two individuals on any plot) were recorded. Phacelia attracted 87–99% of all bee visits over the two years of this study. Buckwheat, a nectar source that comprised 20% of both seed mixtures by weight, attracted 1% or less of all bee visits.

Gathmann & Tscharntke (1997) monitored solitary bees and wasps nesting in reed stem nest boxes placed on three set-aside fields sown with a clover grass mix in Germany over three years. Relative to nest boxes placed in semi-natural grasslands, few species occupied these nest boxes (quantitative details are lacking from the report of this trial).

Carreck *et al.* (1999) recorded 15 species of bee visiting flowers over two summers, in four plots of six annual flowering plant species at Rothamsted Research, Hertfordshire, UK. Short-tongued bumblebees – buff-tailed and red-tailed (*Bombus terrestris/lucorum* and *B. lapidarius/ruderarius*) were the most abundant wild bee visitors, and bees were most numerous on phacelia, borage and (second year only) cornflower *Centaurea cyanus*.

In a replicated study of foraging bee communities on set-aside fields of different ages and management (four replicates of each) in Germany, Steffan-Dewenter & Tscharntke (2001) found that 1-year-old set-aside fields sown with phacelia had a similar abundance but fewer species of bee (13 species/field on average) than 1- to 5-year-old naturally regenerated set-aside fields (15–29 species/field). Bees found on phacelia were mainly common species of bumblebee and the solitary bee genus *Lasioglossum*, whereas several endangered and specialised bees were found foraging on naturally regenerated set-aside.

Carreck & Williams (2002) evaluated a sown mix of six annual flowering species: cornflower, common mallow *Malva sylvestris* (both native), borage, buckwheat, marigold *Calendula officinalis* and phacelia as forage for insects, in four plots at Rothamsted Research, Hertfordshire, UK. The mix attracted 16 bee species, the most numerous insects being honey bee *Apis mellifera* and red-tailed bumblebee *Bombus lapidarius/B. ruderarius* (not distinguished in the study). 97% of all bumblebee visits were to phacelia and borage, 67% of all solitary bee visits were to marigold. The common carder bee *B. pascuorum* and garden bumblebee *B. hortorum* (both common long-tongued species) were recorded in relatively low numbers.

Fluri & Frick (2002) recorded 0.2 bumblebees/m^2 (2,000 bumblebees/ha) foraging on a single 0.3 ha phacelia plot in Switzerland.

Dramstad *et al.* (2003) recorded numbers of bumblebees visiting a single 2 m × 210 m sown strip of phacelia, in Vestby, Norway, in 1994. They recorded a peak of 237 bumblebees on the strip (0.6/m^2) on 17 July, which gradually declined to 93 bumblebees on the strip ($0.2/m^2$) on 28 July.

In a replicated trial of flower-visiting insects foraging on six 0.1 ha sown patches of phacelia on farmland near Thessaloniki, Greece, 95% of all visits were by honey bees (Petanidou 2003). No bumblebees and only small numbers of solitary bees, mostly sweat bees of the family Halictidae (no more than six species in any flowering period, 12 species in total) foraged on the patches.

In a replicated, controlled trial in eastern and central England, Heard *et al.* (2007) showed that patches sown with a 20% legume seed mix (clovers *Trifolium* spp. and bird's-foot trefoil *Lotus corniculatus*) at eight sites attracted significantly higher densities of bumblebees than control patches of non-crop vegetation typical of the site (average 26 bumblebees/200 m^2 on forage patches compared to 2 bumblebees/200 m^2 on control patches). Honey bees *Apis mellifera* and cuckoo bumblebees (*Bombus* [Psithyrus] spp.) were not in greater densities on forage patches. The study also showed that bumblebee densities on sown forage patches were higher in areas with a greater propor-tion of arable land in a surrounding 1 km radius circle of landscape than in landscapes with less arable and more grassland, woodland and urban habi-tat. This demonstrates that planted leguminous forage is more valuable to bumblebees in intensive arable landscapes.

Sheffield *et al.* (2008) demonstrated that planting 3 × 45 m patches of the native bigleaf lupine *Lupinus polyphyllus* in three apple orchards in Novia Sco-tia, Canada, significantly enhanced the reproductive success of managed blue orchard mason bees *Osmia lignaria*. Nests were heavier and contained more new bees/nest in six nest boxes placed next to the lupine patch, relative to six nest boxes placed 600 m away from the lupine patches in the same orchards.

Kohler *et al.* (2008) planted 100 m^2 patches of 17 perennial and annual flowering plant species at five locations on intensive farmland in the central Netherlands. They measured the abundance and diversity of bees during one summer at 10 sampling locations along a 1,500 m transect running away from each patch, and along five 1,500 m control transects. All the transects ran alongside ditches. Bee abundance and diversity were 60–80% higher than on control transects within the flower patches, but not anywhere else along the experimental transects. This suggests that patches of sown forage plants do not enhance numbers of bees in the surrounding landscape, at least in the first year. The lowest values for numbers of bees and bee species were re-corded at the sampling point 50 m away from the flower patches.

Tuell *et al.* (2008) evaluated native perennial plant species in the eastern USA for their attractiveness to wild bees in a replicated experiment (five replicate 1 m² plots of each species). Nine out of 43 species were highly attractive to bees, having an average of five or more wild bees per m² plot in vacuum sampling or timed observation. These were *Potentilla fruticosa, Scrophularia marilandica, Veronicastrum virginicum, Ratibida pinnata, Agastache nepetoides, Silphium perfoliatum, Lobelia siphilitica, Solidago riddellii* and *Solidago speciosa*. Three other plant species (*Zizia aurea, Fragaria virginiana* and *Coreopsis lanceolata*) were identified as attractive to wild bees in the early season (May and June), a crucial time for early-emerging bee species, when flowers are generally less abundant.

A randomised, replicated, controlled trial on four farms in southwest England (Potts *et al.* 2009) found that 50 × 10 m plots of permanent pasture annually sown with a mix of legumes, or grass and legumes, supported more common bumblebees (individuals and species) than seven grass management options. There were twelve replicates of each management, monitored over four years. No more than 2.2 bumblebees/transect were recorded on average on any grassy plot in any year, compared to over 15 bumblebees/transect in both sown treatments in one year. The legumes sown included white clover *Trifolium repens*, red clover *T. pratense*, common vetch *Vicia sativa* and bird's-foot trefoil *Lotus corniculatus*.

Carreck N.L. & Williams I.H. (1997) Observations on two commercial flower mixtures as food sources for beneficial insects in the UK. *Journal of Agricultural Science*, 128, 397–403. www↗

Carreck N.L., Williams I.H. & Oakley J.N. (1999) Enhancing farmland for insect pollinators using flower mixtures. *Aspects of Applied Biology*, 54, 101–108. www↗

Carreck N.L. & Williams I.H. (2002) Food for insect pollinators on farmland: insect visits to flowers of annual seed mixtures. *Journal of Insect Conservation*, 6, 13–23. www↗

Dramstad W.E., Fry G.L.A. & Schaffer M.J. (2003) Bumblebee foraging – is closer really better? *Agriculture, Ecosystems and Environment*, 95, 349–357. www↗

Engels W., Schulz U. & Rädle M. (1994) Use of Tübingen mix for bee pasture in Germany. In: Matheson A. (ed.), *Forage for Bees in an Agricultural Landscape*, pp. 57–65. International Bee Research Association. www↗

Fluri P. & Frick R. (2002) Honey bee losses during mowing of flowering fields. *Bee World*, 83, 109–118. www↗

Gathmann A., Greiler H-J. & Tscharntke T. (1994) Trap-nesting bees and wasps colonizing set-aside fields: succession and body size, management by cutting and sowing. *Oecologia*, 98, 8–14. www↗

Gathmann A. & Tscharntke T. (1997) Bienen und Wespen in der Agrarlandschaft (Hymenoptera Aculeata): Ansiedlung und Vermehrung in Nisthilfen [Bees and wasps in the agricultural landscape (Hymenoptera Aculeata): colonization and augmentation in trap nests]. *Mitteilungen der Deutschen Gesellschaft für allgemeine und angewandte Entomologie*, 11, 91–94. www↗

Heard M.S., Carvell C., Carreck N.L., Rothery P., Osborne J. L. & Bourke A.F.G. (2007) Landscape context not patch size determines bumble-bee density on flower mixtures sown for agri-environment schemes. *Biology Letters*, 3, 638–641. www↗

Kohler F., Verhulst J., van Klink R. & Kleijn D. (2008) At what spatial scale do high-quality habitats enhance the diversity of forbs and pollinators in intensively farmed landscapes? *Journal of Applied Ecology*, 45, 753–762. www↗

Patten K.D., Shanks C.H. & Mayer D.F. (1993) Evaluation of herbaceous plants for attractiveness to bumble bees for use near cranberry farms. *Journal of Apicultural Research*, 32, 73–79. www↗

Petanidou T. (2003) Introducing plants for bee-keeping at any cost? – Assessment of *Phacelia tanacetifolia* as nectar source plant under xeric Mediterranean conditions. *Plant Systematics and Evolution*, 238, 155–168. www↗

Potts S.G., Woodcock B.A., Roberts S.P.M., Tscheulin T., Pilgrim E.S., Brown V.K. & Tallowin J.R. (2009) Enhancing pollinator biodiversity in intensive grasslands. *Journal of Applied Ecology*, 46, 369–379. www↗

Sheffield C.S., Westby S.M., Smith R.F. & Kevan P.G. (2008) Potential of bigleaf lupine for building and sustaining *Osmia lignaria* populations for pollination of apple. *The Canadian Entomologist*, 140, 589–599. www↗

Steffan-Dewenter I. & Tscharntke T. (2001) Succession of bee communities on fallows. *Ecography*, 24, 83–93. www↗

Tuell, J.K., Fiedler A.K., Landis D. & Isaacs R. (2008). Visitation by wild and managed bees (Hymenoptera : Apoidea) to eastern US native plants for use in conservation programs. *Environmental Entomology*, 37, 707–718. www↗

Williams I.H & Christian D.G. (1991) Observations on *Phacelia tanacetifolia* Bentham (Hydrophyllaceae) as a food plant for honey bees and bumble bees. *Journal of Apicultural Research*, 30, 3–12. www↗

3.14 Sow uncropped arable field margins with an agricultural 'nectar and pollen' mix

See also Chapter 3.13 *Plant dedicated floral resources on farmland.*

• Five replicated trials in Europe (three controlled) have documented bumblebees foraging on field margins sown with an agricultural nectar and pollen seed mix. Four replicated trials showed that field margins sown with perennial leguminous flowering plants attract significantly more foraging bumblebees than naturally re-generated (two trials), grassy (four trials) or cropped (three trials) field margins. Three replicated trials showed that a mix of agricultural forage plants including legumes (all annual plants in one trial) attracts greater numbers of bumblebees than a perennial wildflower mix, at least in the first year.

• Three trials in the UK found evidence that margins sown with agricultural legume plants degrade in their value to bumblebees and would need to be re-sown every few years.

• We have captured no evidence on the effects of field margin management on solitary bees.

Arable field margins are the focus of some specific European agri-environment measures. This section covers 'nectar and pollen' seed mixes comprising agricultural forage plants or non-native plants, designed to provide nectar and pollen sources for bees and other flower-visiting insects. For nectar and pollen mixes exclusively composed of native wild flowers, see Chapter 3.15 *Sow uncropped arable field margins with a native wild flower seed mix.*

In a replicated trial in central Sweden (Lagerlöf *et al.* 1992), four 2 m wide field margins sown with a mix of legumes dominated by either red clover *Trifolium pratense* or white melilot *Melilota alba* attracted significantly more bumblebees *Bombus* spp. than naturally regenerated field margins and species-rich dry pasture. Red clover was the most attractive, with 72% of a total of 413 individual bees recorded on that treatment. White melilot was extremely attractive to honey bees *Apis mellifera*, attracting 98% of the 2,422 recorded in the study.

In a replicated trial (five plots) of field margin seed mixtures on a farm in North Yorkshire, Carvell *et al.* (2006) found that short-tongued bumblebees (*B. terrestris, B. lucorum* and *B. pratorum*) strongly preferred plots of annually sown cover crops including borage *Borago officinalis* and common melilot *Me-*

lilotus officinalis over perennial wildflower seed mix. Total bumblebee abundance was higher on the annual agricultural nectar mix. On average 70% of pollen collected by buff-tailed bumblebee workers *B. terrestris* sampled in this study was from borage.

In a replicated controlled trial in thirty-two 10 km grid squares across England (Pywell *et al.* 2006), there were significantly more bumblebee species and more individuals on field margins sown 1–2 years previously with a pollen and nectar mix (average >3 species and 86 bees/transect) than on grassy margins (average 1.3–1.4 species and 6–8 bees/transect) or cropped margins (average 0.1 species and 0.2 bees/transect). There were more bumblebee individuals, but not more bumblebee species on pollen and nectar mix margins (average 86 bees/transect) than on wildflower-sown margins (43 bees/transect). The abundance of long-tongued bumblebees (mostly common carder bee *B. pascuorum* and garden bumblebee *B. hortorum*) was positively correlated with the number of pollen and nectar-mix agreements in each 10 km square.

In a replicated controlled trial at six sites across central and eastern England, Carvell *et al.* (2007) found that 6 m margins of cereal fields sown with a nectar flower mixture supported significantly more foraging bumblebees (species and individuals) than cropped, grassy or naturally regenerated unsown field margins. Visitors included the nationally rare long-tongued species *Bombus ruderatus* and *B. muscorum*.

The nectar flower seed mixture was based on four agricultural legumes (red clover, Alsike clover *Trifolium hybridum,* bird's-foot trefoil *Lotus corniculatus* and sainfoin *Onobrychis viciifolia*). Unlike a wild flower seed mixture in the same study, it supported more bumblebees than other treatments from the first year of the study. However, relative to a wild flower mixture, this mixture provided low numbers of flowers in May and June, when bumblebee queens of late-emerging species are foraging. It also showed a decline in flower numbers in year three, when it did not support significantly more bumblebees than the wild flower seed mixtures.

Arable margins sown with legume-grass seed mix had higher species richness of bumblebee forage plants (almost 100% cover of Alsike clover and red clover one year after establishment) over four years, compared to naturally regenerated margins on farmland at Romney Marsh, Kent, UK (Gardiner *et al.* 2008). Bee visits were not reported in this study. Fixed-time transect walks in the clover margins are reported elsewhere (Edwards & Williams 2004) to have demonstrated a 300-fold increase in bumblebee forager numbers in the margins planted with clover, but unfortunately, no control count was carried out for comparison. However, the clover-sown plots were invaded by perennial grasses in the third and fourth years of this study, and flower numbers fell substantially.

In a replicated, controlled study (2005–2008), across 41 farms in England and Scotland, the average number of worker bumblebees was greater on margins where legume-rich seed mix was established than on other field margins (grassy margins or track edges; Edwards 2008). No formal statistical analysis were performed on these data. There was an observed decline in the relative number of foraging worker bumblebees on these margins after they had been established for more than three years (data from five farms).

Carvell C., Meek W.R., Pywell R.F., Goulson D. & Nowakowski M. (2007) Comparing the efficacy of agri-environment schemes to enhance bumble bee abundance and diversity on arable field margins. Journal of Applied Ecology, 44, 29–40. www↗

Carvell C., Westrich P., Meek W.R., Pywell R.F. & Nowakowski M. (2006) Assessing the value of annual and perennial forage mixtures for bumblebees by direct observation and pollen analysis. Apidologie, 37, 326–340. www↗

Edwards M. (2008) Syngenta Operation Bumblebee Monitoring Report 2005-2008. Report to Syngenta. www↗

Edwards M. & Williams P.H. (2004) Where have all the bumblebees gone, and could they ever return? British Wildlife, 15, 305–312.

Gardiner T., Edwards M. & Hill J. (2008) Establishment of clover-rich field margins as a forage resource for bumblebees Bombus spp. on Romney Marsh, Kent, England. Conservation Evidence, 5, 51–57. www↗

Lagerlöf J., Stark J. & Svensson B. (1992) Margins of agricultural fields as habitats for pollinating insects. Agriculture, Ecosystems and Environment, 40, 117–124. www↗

Pywell R.F., Warman E.A., Hulmes L., Hulmes S., Nuttall P., Sparks T.H., Critchley C.N.R. & Sherwood A. (2006) Effectiveness of new agri-environment schemes in providing foraging resources for bumblebees in intensively farmed landscapes. Biological Conservation, 129, 192–206. www↗

3.15 Sow uncropped arable field margins with a native wild flower seed mix

- Five replicated trials in the UK showed that uncropped field margins sown with wild flowers and subsequently mown support a higher abundance (and in three trials higher species richness) of foraging bumblebees than cropped field edges (all five trials), grassy margins (four trials) or naturally regenerated uncropped margins (three trials). One small trial recorded the same number of bee species on wildflower sown and naturally regenerated strips.

- Two trials demonstrated that perennial leguminous herbs in the seed mixtures are important forage sources for bumblebees, particularly for long-tongued species.

- One small replicated trial showed that common long-tongued bumblebee species (*Bombus pascuorum* and *B. hortorum*) strongly preferred plots of perennial wild-flower seed mix over a mix of annual forage plants.

- We have captured no evidence on the effects of field margin management on solitary bees.

Nine bee species were recorded on three field margin strips sown with a diverse grass and wildflower seed mix established for three years at the ADAS Bridgets Research Centre, Hampshire, UK in 1998 (Carreck *et al.* 1999). The same number of species was recorded on a single naturally regenerated field margin strip in the same study.

A small-scale replicated, controlled trial of field margin management options on two farms in North Yorkshire, UK in one summer (Meek *et al.* 2002) found a significantly greater abundance of bumblebees *Bombus* spp. on four 6 m wide margins sown with a grass and wildflower seed mix than on four naturally regenerated, grass-sown or control cropped margins.

In a replicated controlled three-year trial on three arable field margins at a farm in North Yorkshire, Carvell *et al.* (2004) found 6 m wide field margin plots sown, or half-sown with a native 'grass and wildflower' seed mix supported significantly more bumblebees than margins sown with tussocky grass, or control cropped field margins. The wildflower-sown margins supported significantly more bumblebees than naturally regenerated margins in the same experiment, in the first year of the study only, and this difference was not significant when data were averaged across all three years. However, the margins sown with wildflower seed mix supported consistently high numbers of bumblebees, whereas the naturally regenerated margins had one bumper year for bumblebees and were poor in the other two years.

In a replicated controlled trial in central and eastern England (Pywell *et al.* 2005), bumblebee foraging activity and species richness were significantly enhanced at 28 uncropped field margins sown with a 'wildlife seed mixture', compared to paired control sites of conventionally managed cereal or 16 'conservation headlands'. The seed mixture contained grasses, and annual and perennial broad-leaved herbs. This result was dependent upon key forage species being included in the seed mixture, including red clover *Trifolium pratense*, bird's-foot trefoil *Lotus corniculatus* and borage *Borago officinalis*, the latter being of particular importance to short-tongued bumblebee species such as *Bombus terrestris* and *B. lucorum*.

In a replicated trial (five plots) of field margin seed mixtures on a farm in North Yorkshire, Carvell *et al.* (2006) found that both common UK species of long-tongued bumblebee (*B. pascuorum* and *B. hortorum*) strongly preferred plots of perennial wildflower seed mix over annually sown agricultural nectar plants (including borage *Borago officinalis*, fodder radish *Raphanus sativus* and common melilot *Melilotus officinalis*), although total bumblebee abundance was highest on the annual agricultural nectar mix. On average 76% of pollen collected by common carder bee *B. pascuorum* workers sampled in this study was from red clover.

In a replicated controlled trial in thirty-two 10 km grid squares across England (Pywell *et al.* 2006), there were significantly more bumblebee species and more individuals on field margins sown with a wildflower seed mix (average >3 species and 43 bees/transect) than on grassy margins (average 1.3–1.4 species and 6–8 bees/transect) or cropped margins (average 0.1 species and 0.2 bees/transect).

In a replicated controlled trial at six sites across central and eastern England (Carvell *et al.* 2007), 6 m margins of cereal fields sown with 21 annual and perennial wild flowers supported significantly more foraging bumblebees (species and individuals) than cropped field margins (including conservation headlands). In years two and three of the study, these seed mixtures also supported more foraging bumblebees (species and individuals) than grassy or naturally regenerated unsown field margins, and in the third year they supported as many bees as a nectar mix based on agricultural legumes. The wildflower mixture took a year to establish properly, but may provide forage for a longer period of the year than the nectar mix. It is also likely to persist for five to 10 years, not declining in flower numbers after three years like the nectar mix.

Carreck N.L., Williams I.H. & Oakley J.N. (1999) Enhancing farmland for insect pollinators using flower mixtures. *Aspects of Applied Biology*, 54, 101–108. www↗

Carvell C., Meek W.R., Pywell R.F. & Nowakowski M. (2004) The response of bumblebees to successional change in newly created arable field margins. *Biological Conservation*, 118, 327–339. www↗

Carvell C., Meek W.R., Pywell R.F., Goulson D. & Nowakowski M. (2007). Comparing the efficacy of agri-environment schemes to enhance bumble bee abundance and diversity on arable field margins. *Journal of Applied Ecology*, 44, 29–40. www↗

Carvell C., Westrich P., Meek W.R., Pywell R.F. & Nowakowski M. (2006) Assessing the value of annual and perennial forage mixtures for bumblebees by direct observation and pollen analysis. *Apidologie*, 37, 326–340. www↗

Meek B., Loxton D., Sparks T., Pywell R., Pickett H. & Nowakowski M. (2002) The effect of arable field margin composition on invertebrate biodiversity. *Biological Conservation*, 106, 259–271. www↗

Pywell R.F., Warman E.A., Carvell C., Sparks T.H., Dicks L.V., Bennett D., Wright A., Critchley C.N.R. & Sherwood A. (2005) Providing foraging resources for bumblebees in intensively farmed landscapes. *Biological Conservation*, 121, 479–494. www↗

Pywell R.F., Warman E.A., Hulmes L., Hulmes S., Nuttall P., Sparks T.H., Critchley C.N.R. & Sherwood A. (2006) Effectiveness of new agri-environment schemes in providing foraging resources for bumblebees in intensively farmed landscapes. *Biological Conservation*, 129, 192–206. www↗

3.16 Leave arable field margins uncropped with natural regeneration

• Four replicated trials in the UK have found more bumblebees (and more bee species in two trials) foraging on uncropped field margins than on cropped margins. One small unreplicated trial found similar bee species richness on a naturally regenerated margin as on margins sown with wildflowers. A small replicated trial found that neither abundance nor diversity of bumblebees were higher on naturally regenerated margins than on cropped margins.

• Two trials note that the value of naturally regenerated uncropped field margins is based on thistle species considered to be pernicious weeds requiring control. Two trials found that the value of naturally regenerated uncropped field margins for bees was not consistent from year to year.

• We have captured no evidence on the effects of field margin management on solitary bees.

Nine bee species were recorded on a single naturally regenerated field margin strip established for three years at ADAS Bridgets Research Centre, Hampshire, UK in 1998 (Carreck *et al.* 1999), the same number of species as on three strips sown with a diverse wildflower seed mix in the same study.

A small replicated, controlled trial of field margin management options on two farms in North Yorkshire, UK in one summer (Meek *et al.* 2002) did not find significantly more bumblebees *Bombus* spp. (species or individuals) on four naturally regenerated 6 m wide margins than on four cropped margins.

A replicated controlled trial of UK arable field margins allowed to regenerate naturally for one year found that they supported significantly more honey bees and bumblebees than unsprayed cropped margins managed as conservation headlands (averages between 10 and 50 bees/transect on naturally regenerated margins compared to <3 bees/transect in conservation headlands; Kells *et al.* 2001).

In a replicated controlled trial on three arable field margins at one farm in North Yorkshire, Carvell *et al.* (2004) found 6 m wide naturally regenerated, uncultivated field margin plots supported significantly more foraging bumblebees than margins sown with tussocky grass, or control cropped field margins, but only in one year (2001) of this three year study. In the other two years (2000 and 2002), the naturally regenerated field margins did not support significantly more bumblebees than the control or grass-sown sites. In 2001, the bumblebees were mostly foraging on spear thistle *Cirsium vulgare*, a pernicious agricultural weed that had to be controlled by cutting at the end of that summer. Naturally regenerated margins were the only treatment that did not support consistent numbers of bumblebees in all three years.

Bumblebee foraging activity and species richness were significantly enhanced on 18 uncropped, regularly cultivated field margins where natural regeneration had been allowed to take place for five years, compared to paired control sites of conventionally managed cereal, in East Anglia and the West Midlands, UK (Pywell *et al.* 2005). The uncropped margins had significantly more plant species than either conservation headlands or uncropped margins sown with a wildflower seed mix. However, two species considered to be pernicious weeds, spear thistle and creeping thistle *C. arvense* were key forage plants for the bumblebees, so this option may lead to conflict between agricultural and conservation objectives.

The naturally regenerated field margins supported fewer bumblebees (18 individuals and 2.7 species/100 m on average) than margins sown with a wild flower seed mixture (29 bumblebees, 3.0 species/100 m), but the two treatments were not directly compared in the analysis.

In a replicated controlled trial at six sites across central and eastern England, Carvell *et al.* (2007) found that naturally regenerated field margins supported a greater number and diversity of foraging bumblebees than cropped margins (including conservation headlands), but only in the first year of the study. In subsequent second and third years, bumblebee numbers were not significantly different from cropped treatments, but this may be due to the presence of more attractive floral resources planted on the same field margins for the experiment.

Carreck N.L., Williams I.H. & Oakley J.N. (1999) Enhancing farmland for insect pollinators using flower mixtures. *Aspects of Applied Biology*, 54, 101–108. www↗

Carvell C., Meek W.R., Pywell R.F., Goulson D. & Nowakowski M. (2007) Comparing the efficacy of agri-environment schemes to enhance bumble bee abundance and diversity on arable field margins. *Journal of Applied Ecology*, 44, 29–40. www↗

Carvell C., Meek W.R., Pywell R.F. & Nowakowski M. (2004) The response of bumblebees to successional change in newly created arable field margins. *Biological Conservation*, 118, 327–339. www↗

Kells A.R., Holland J. & Goulson D. (2001) The value of uncropped field margins for foraging bumblebees. *Journal of Insect Conservation*, 5, 283–291. www↗

Meek B., Loxton D., Sparks T., Pywell R., Pickett H. & Nowakowski M. (2002) The effect of arable field margin composition on invertebrate biodiversity. *Biological Conservation*, 106, 259–271. www↗

Pywell R.F., Warman E.A., Carvell C., Sparks T.H., Dicks L.V., Bennett D., Wright A., Critchley C.N.R. & Sherwood A. (2005) Providing foraging resources for bumblebees in intensively farmed landscapes. *Biological Conservation*, 121, 479–494. www↗

3.17 Increase the diversity of nectar and pollen plants in the landscape

- One large replicated controlled trial showed that the average abundance of long-tongued bumblebees on field margins was positively correlated with the number of 'pollen and nectar' agri-environment agreements in a 10 km grid square.

Background

Managing landscapes to enhance nectar and pollen resources for bees throughout their life cycle is increasingly recognised as an important strategy to enhance the agricultural pollination service and to conserve bee populations. It could involve increasing the diversity of flowering crops or conserving aspects of landscape, such as woodlands or riparian areas, which provide floral resources in dry or early spring periods.

For example, using a diversity of flowering shade trees over crops such as cardamom and coffee in India is recommended to encourage the giant honey bee *Apis dorsata* not to migrate, potentially enhancing local honey bee populations (FAO 2008).

Gemmill-Herren & Ochieng (2008) demonstrate the importance of riparian woodland at the height of the dry season to wild pollinators of egg plant *Solanum melongena* in southwest Kenya. In temperate regions, early spring flowers provide particularly crucial resources to queen bumblebees.

In Europe, recent research has shown that higher coverage of the mass-flowering crop oilseed rape *Brassica napus* in the landscape is associated with higher numbers of foraging worker bumblebees at focal sampling points, but not with enhanced bumblebee reproductive success or colony densities in Europe (Westphal *et al.* 2003, Hermann *et al.* 2007, Westphal *et al.* 2009). This work is not summarised on Conservation Evidence because increasing the coverage of one particular flowering crop is not yet considered a conservation intervention. This may change in future, as knowledge develops.

We have captured no direct evidence on the effects of managing elements of landscape such as areas of natural or semi-natural habitat, or crop diversity, to enhance nectar and pollen resources. One piece of evidence demonstrates that enhancing the quantity of planted nectar and pollen resources at the landscape (10 km²) scale benefits bees.

In a replicated controlled trial in thirty-two 10 km grid squares across England (Pywell *et al.* 2006), the abundance of long-tongued bumblebees, mostly common carder bee *B. pascuorum* and garden bumblebee *B. hortorum*, recorded on trial field margins (various planting treatments) was positively correlated with the total number of pollen and nectar-mix agri-environment agreements in each 10 km square. There is no record of the numbers of long-tongued bumblebees in these grid squares before the agreements were implemented.

Food and Agriculture Organization of the United Nations (FAO) (2008) *Initial survey of good pollination practices.* FAO, Rome. Available at: http://www.internationalpollinatorsinitiative.org/uploads/SURVEY%20DEC%2008%20Small.pdf.

Gemmill-Herren B. & Ochieng A.O. (2008) Role of native bees and natural habitats in eggplant (*Solanum melongena*) pollination in Kenya. *Agriculture, Ecosystems and Environment*, 127, 31–36.

Herrmann F., Westphal C., Moritz R.F.A. & Steffan-Dewenter I. (2007) Genetic diversity and mass resources promote colony size and forager densities of a social bee (*Bombus pascuorum*) in agricultural landscapes. *Molecular Ecology*, 16, 1167–1178.

Pywell R.F., Warman E.A., Hulmes L., Hulmes S., Nuttall P., Sparks T.H., Critchley C.N.R. & Sherwood A. (2006) Effectiveness of new agri-environment schemes in providing foraging resources for bumblebees in intensively farmed landscapes. *Biological Conservation*, 129, 192–206. www↗

Westphal C., Steffan-Dewenter I. & Tscharntke T. (2003) Mass flowering crops enhance pollinator densities at a landscape scale. *Ecology Letters*, 6, 961–965.

Westphal C., Steffan-Dewenter I. & Tscharntke T. (2009) Mass flowering oilseed rape improves early colony growth but not sexual reproduction of bumblebees. *Journal of Applied Ecology*, 46, 187–193.

3.18 Reduce the intensity of farmland meadow management

See also Chapter 4.1 *Introduce agri-environment schemes that reduce spraying.*

- Four replicated trials in Europe have compared farmland meadows managed extensively with conventionally farmed meadows or silage fields. Two found enhanced numbers and diversity of wild bees on meadows with a delayed first cut and little agrochemical use. Two found no difference in bee diversity or abundance between conventional meadows and meadows with reduced fertilizer use or cutting intensity.

Reducing the intensity of grassland management involves reducing or stopping the use of fertilizers, herbicides and pesticides and delaying the mowing date until later in the summer.

In a replicated trial in Switzerland (Knop *et al.* 2006), 21 hay meadows managed under the 'Ecological Compensation Areas' agri-environment scheme for three to 10 years had more species of wild bee and more individual wild bees than 21 paired conventionally managed hay meadows. There were 13 species/field, 11 individuals/survey under the agri-environment schemes, compared to 11 species/field and 8 individuals/survey on conventional meadows. This agri-environment scheme requires a postponed first cut, in June or later, and no additions of fertilizer or pesticide to the meadow, although in the study three of the trial meadows were fertilized a little, despite the regulations.

In a similar replicated trial in the Netherlands (Kohler *et al.* 2007), an agri-environment scheme aimed at enhancing habitat for birds by reducing fertilizer and pesticide input and delaying cutting or grazing had no impact on diversity or numbers of non-*Apis* bees in 21 Dutch wet meadow fields when compared with paired conventionally managed fields. Bee diversity and abundance was low in both field types (average <3 species/field; <6 individuals per field). However, this agri-environment scheme allowed application of nitrogen fertilizer at 206 kg/ha, which is 75% of the standard fertilizer application rate (269 kg/ha). The meadows had been under the scheme for between three and 10 years.

A replicated trial of 13 meadows under the Swiss Ecological Compensation Area agri-environment scheme in 2004 found that the species richness and abundance of solitary and social bees visiting potted flowering plants were higher in meadows under the scheme than in adjacent, intensively managed meadows (Albrecht *et al.* 2007).

A randomised, replicated, controlled trial on four farms in southwest England (Potts *et al.* 2009) found that 50 × 10 m plots of permanent pasture managed more extensively without fertilizer or without grazing, and/or with a higher cutting height or reduced cutting frequency did not support more common bumblebees *Bombus* spp. than control plots conventionally managed for silage. There were twelve replicates of each management type, monitored over four years. No more than 2.2 bumblebees/transect were recorded on average on any grassy plot in any year.

A systematic review is currently underway to assess the effect on biodiversity of delayed first mowing date on annually mown hay meadows in Europe (Pellet & Wunderlin, in prep).

Albrecht M., Duelli P., Müller C., Kleijn D. & Schmid B. (2007) The Swiss agri-environment scheme enhances pollinator diversity and plant reproductive success in nearby intensively managed farmland. *Journal of Applied Ecology*, 44, 813–822. www↗

Knop E., Kleijn D., Herzog F. & Schmid B. (2006) Effectiveness of the Swiss agri-environment scheme in promoting biodiversity. *Journal of Applied Ecology*, 43, 120–127. www↗

Kohler F., Verhulst J., Knop E., Herzog F. & Kleijn D. (2007) Indirect effects of grassland extensification schemes on pollinators in two contrasting European countries. *Biological Conservation*, 135, 302–307. www↗

Pellet J. & Wunderlin J. (in prep). *Does delaying the first mowing date increase biodiversity in European farmland meadows?* Systematic Review no. 72. Collaboration for Environmental Evidence. http://www.environmentalevidence.org/SR72.html

Potts S.G., Woodcock B.A., Roberts S.P.M., Tscheulin T., Pilgrim E.S., Brown V.K. & Tallowin J.R. (2009) Enhancing pollinator biodiversity in intensive grasslands. *Journal of Applied Ecology*, 46, 369–379. www↗

3.19 Reduce grazing intensity on pastures

• One replicated trial has shown that reducing the intensity of summer cattle grazing can increase the abundance, but not the species richness of cavity-nesting bees and wasps.

Background

The effects of grazing on wild bee communities have been studied in a number of cases by examining how wild bees are distributed in habitats with different historic grazing regimes. The effects of grazing seem to be different in different contexts. Studies have found lower numbers of wild bee species in grazed areas, compared to ungrazed areas (Hatfield & LeBruhn 2007,) lower abundance and diversity of wild bees after severe grazing by deer (Nakamura & Ono 1999), lower abundance of bees but not lower bee species richness in grazed areas (Kearns & Oliveras 2009) and higher species richness and abundance of bumblebees *Bombus* sp. in cattle-grazed areas compared to ungrazed areas (Carvell 2002). Two studies have found no difference in bee abundance or species richness between grazed and ungrazed orchard meadows (Steffan-Dewenter & Leschke 2003), or intensively and extensively grazed grasslands (Sarospataki *et al.* 2009).

These studies have not directly reduced the intensity of grazing, so they do not represent evidence on the effects of this action for bee conservation.

In a comparison of six intensively (5.5 cattle/ha) and six lightly (1.5 cattle/ha) cattle-grazed meadows with six ungrazed meadows in Germany, meadows with light grazing had a greater number of individual cavity-nesting bees, wasps and their brood parasites than meadows with intensive grazing (Kruess & Tscharntke 2002). There was an average of 47 emerging individuals/lightly grazed site, compared to 27 emerging individuals/intensively grazed site. Reduced intensity of grazing did not significantly increase the number of bee and wasp species.

Both abundance and total species richness of these insects were significantly higher on ungrazed grassland (11.5 species) than on intensively

(4.7 species) or lightly (6.2 species) grazed pastures. These results were linked to an increase in vegetation height as grazing intensity is reduced.

Carvell C. (2002) Habitat use and conservation of bumblebees (*Bombus* spp.) under different grassland management regimes. *Biological Conservation*, 103, 33–49.

Hatfield R.G. & LeBuhn G. (2007) Patch and landscape factors shape community assemblage of bumble bees, *Bombus* spp. (Hymenoptera: Apidae), in montane meadows. *Biological Conservation*, 139, 150–158.

Kearns C.A. & Oliveras D.M. (2009) Environmental factors affecting bee diversity in urban and remote grassland plots in Boulder, Colorado. *Journal of Insect Conservation*, 13, 655–665.

Kruess A. & Tscharntke T. (2002) Grazing intensity and the diversity of grasshoppers, butterflies, and trap-nesting bees and wasps. *Conservation Biology*, 16, 1570–1580. www↗

Nakamura K. & Ono T. (1999) Influence of deer grazing on the wild bee population in Nikko. *Bulletin of the College of Agriculture Utsunomiya University*, 17, 1–8.

Sarospataki M., Baldi A., Batary P., Jozan Z., Erdos S. & Redei T. (2009) Factors affecting the structure of bee assemblages in extensively and intensively grazed grasslands in Hungary. *Community Ecology*, 10, 182–188.

Steffan-Dewenter I. & Leschke K. (2003) Effects of habitat management on vegetation and above-ground nesting bees and wasps of orchard meadows in Central Europe. *Biodiversity and Conservation*, 12, 1953–1968.

4. Threat: pollution - agricultural and forestry effluents

Introduce agri-environment schemes that reduce spraying

Four replicated trials in Europe have shown enhanced diversity and/or abundance of foraging wild bees on land managed under various European agri-environment schemes, relative to conventional fields or field margins. Four replicated trials found that the number of bees and/or bee species is not enhanced on land managed under European agri-environment schemes. On a landscape scale, two replicated trials in the UK have found bumblebee populations were not enhanced in areas with farmland managed under agri-environment schemes.

Convert to organic farming

Six replicated trials from Europe or North America have monitored bees on organic and conventional arable farms. Three trials showed that the abundance of wild bees is higher under organic farming. Three trials found no significant difference in the numbers of bumblebees (two trials), bumblebee species (one trial), or wild bees visiting flowering crops (one trial) between conventional and organic farms.

Restrict the use of certain pesticides

One site comparison study in Italy showed that a reduction in the number of solitary bee species in late summer can be avoided by not applying the insecticide fenitrothion repeatedly.

Reduce pesticide or herbicide use generally

One replicated trial in the USA showed that numbers of foraging bees on squash farms are not affected by the responsible use of pesticides.

Reduce fertilizer run-off into field margins

We have captured no evidence on the effects of specific interventions for reducing fertilizer run-off.

Leave field margins unsprayed within the crop ('conservation headlands')

Two replicated controlled trials in the UK have shown that conservation headlands do not attract more foraging bumblebees than conventional cropped field margins.

Background

The interventions that form the first two sections of this chapter – *Introduce agri-environment schemes that reduce spraying* and *Convert to organic farming* – are placed here because they have a strong component of reduced chemical use, or because the studies monitored effects of a number of different agri-environment schemes, all of which reduced agrochemical inputs. Clearly these interventions also alter aspects of landscape and habitat so their effects cannot be entirely attributed to the change in chemical use. Where a very specific change of land use or habitat is included as part of these schemes, the evidence is repeated in the relevant section in Chapter 3 *Land use change due to agriculture*.

4.1 Introduce agri-environment schemes that reduce spraying

- Four replicated trials in Europe have shown enhanced diversity and/or abundance of foraging wild bees on land managed under various European agri-environment schemes, relative to conventionally-managed fields or field margins. These schemes were the Swiss Ecological Compensation Areas (one replicated trial), the German organic arable farming option (one replicated trial), the Dutch botanical and meadow bird agreements (one replicated trial, very low numbers of bee species) and the Scottish Rural Stewardship Scheme (one replicated trial, also included nest-searching queen bumblebees).

- Four replicated trials in Europe found that the number of bees and/or bee species is not enhanced on land managed under agri-environment schemes, including meadow bird agreements in wet grassland in the Netherlands, measures to protect steppe-living birds and compensation measures around a National Park in Spain, and 6 m wide grass field margin strips in England (one replicated trial for each).

- On a wider landscape scale, two replicated trials in the UK have found bumblebee populations were not enhanced on farmland managed under agri-environment schemes. One trial compared the reproductive success of colonies of the buff-tailed bumblebee *Bombus terrestris*, the other compared queen bumblebee numbers in spring in conventionally managed field margins, on farms with and without agri-environment schemes.

Background

This section covers studies or reviews that examine the impacts of a range of agri-environment schemes, with reduced agrochemical use in common. For evidence relating to specific agri-environment prescriptions, see also the following sections Chapter 3 *Land Use Change Due to Agriculture*: 3.8 *Increase areas of rough grassland for bumblebee nesting*, 3.10 *Provide grass strips at field margins*, 3.11 *Manage hedges to benefit bees*, 3.14 *Sow uncropped arable field margins with an agricultural nectar and pollen mix*, 3.15 *Sow uncropped arable field margins with a native wild flower seed mix*, 3.18 *Reduce the intensity of farmland meadow management* and 3.19 *Reduce grazing intensity on pastures*. Some of the studies included here are also included under the relevant specific sections.

In a replicated trial with 39 pairs of fields, Kleijn *et al.* (2001) found meadow bird agreements and/or botanical agreements, aimed at conserving wading birds and species-rich vegetation, respectively, in the Netherlands, enhanced the number of bee species relative to conventionally managed control fields. Bee diversity was very low in this study, sampled using 15-minute transect walks (not sweep nets). Three species – honey bee *Apis mellifera*, common carder bee *Bombus pascuorum* and buff-tailed bumblebee *B. terrestris* – accounted for 85% of bees recorded.

Goulson *et al.* (2002) compared the growth of experimental *B. terrestris* colonies placed on 10 farms with substantial conservation measures with those placed on 10 conventional arable farms. Conservation measures included conservation headlands, set-aside and minimal use of pesticides. This study found no measurable difference between colonies on the different types of farm. The authors suggest this is because *B. terrestris* has a foraging range that extends beyond individual farms, which may not be true for other bumblebee species.

Kleijn & Sutherland (2003) reviewed studies of the effectiveness of European agri-environment schemes in published and unpublished literature. Three out of the 62 studies included bees. Two studies (Kleijn *et al.* 2001, reported above, and Allen *et al.* 2001) found more bees (more species of bee in the case of Kleijn *et al.* 2001) on agri-environment fields compared to control fields under certain schemes. The third study (Kleijn *et al.* 1999) is not reported to have found a difference in bee abundance or species richness between seven agri-environment fields and seven control fields.

A replicated controlled trial of agri-environment schemes on 21 pairs of fields in each of five European countries carried out in 2003 found significantly greater abundance and diversity of wild bees on fields managed under agri-environment schemes than on control fields in Germany and Switzerland, but no significant difference in the Netherlands, England or Spain (Kleijn *et al.* 2006). The agri-environment management options that benefited bees in this study were organic arable farming in Germany and Ecological Compensation Areas in Switzerland. Those that did not were meadow bird agreements in wet grassland in the Netherlands (bees sampled with sweep nets and transect walks), measures to protect steppe-living birds and compensation measures around Cabañeros National Park in Spain, and 6 m wide grass field margin strips in England.

In a replicated controlled trial involving 10 farms in east and central Scotland, Lye *et al.* (2009) compared numbers of nest-searching and foraging queen bumblebees *Bombus* spp. on land that had been managed under three different options of the 'Rural Stewardship Scheme' (unsprayed grassy field margins, species-rich grassland and hedgerows) for three years with conventionally managed land of the same type. On farms with the agri-environment scheme, transects under the scheme attracted significantly more nest-searching and foraging queen bumblebees than conventionally managed transects. However, on conventionally managed transects, there was no significant difference between farms with and without agri-environment schemes in numbers of nest-searching queens, and conventionally managed farms had more foraging queens.

Allen D.S., Gundrey A.L. & Gardner S.M. (2001) *Bumblebees. Technical appendix to ecological evaluation of arable stewardship pilot scheme 1998–2000*. ADAS, Wolverhampton, UK.

Goulson D., Hughes W.O.H., Derwent L.C. & Stout J.C. (2002) Colony growth of the bumblebee, Bombus terrestris, in improved and conventional agricultural and suburban habitats. *Oecologia*, 130, 267–273. www↗

Kleijn D., Boekhoff M., Ottburg F., Gleichman M. & Berendse F. (1999) De effectiviteit van agrarisch natuurbeheer. *Landschap*, 16, 227–235.

Kleijn D., Baquero R.A., Clough Y., Diaz M., De Esteban J., Fernandez F., Gabriel D., Herzog F., Holzschuh A., Johl R., Knop E., Kruess A., Marshall E.J.P., Steffan-Dewenter I., Tscharntke T., Verhulst J., West T.M. & Yela J.L. (2006) Mixed biodiversity benefits of agri-environment schemes in five European countries. *Ecology Letters*, 9, 243–254. www↗

Kleijn D., Berendse F., Smit R. & Gilissen N. (2001) Agri-environment schemes do not effectively protect biodiversity in Dutch agricultural landscapes. Nature, 413, 723–725. www↗

Kleijn D. & Sutherland W.J. (2003) How effective are European agri-environment schemes in conserving and promoting biodiversity? *Journal of Applied Ecology*, 40, 947–969. www↗

Lye G., Park K., Osborne J., Holland J. & Goulson D. (2009) Assessing the value of Rural Stewardship schemes for providing foraging resources and nesting habitat for bumblebee queens (Hymenoptera: Apidae). *Biological Conservation*, 142, 2023–2032. www↗

4.2 Convert to organic farming

• Evidence on the impact of organic farming on wild bees is equivocal. Three replicated trials in Europe or Canada have shown that the abundance of wild bees is higher under organic arable farming than under conventional farming. One of these showed that bee diversity is higher in organically farmed wheat fields and in mown fallow strips adjacent to them. Three replicated trials in Europe or the USA have found no significant difference in the numbers of bumblebees (two trials), bumblebee species (one trial), or wild bees visiting flowering crops (one trial) between conventional and organic arable farms.

Background

Organic farming is supported as a measure to conserve biodiversity under European agri-environment schemes. It disallows the use of mineral fertilisers and synthetic pesticides and herbicides. The soil is kept fertile with regular use of organic manures and nitrogen-fixing leguminous plants. Pest and weed control are achieved through crop rotation, mechanical weeding and inter-cropping.

Belfrage *et al.* (2005) counted bumblebees *Bombus* spp. on six organic and six conventional arable farms in Roslagen, southeastern Sweden. They found no significant difference in the numbers of bumblebees between the two farm types.

A comparison of organic and conventional canola (oilseed rape *Brasscia* sp.) fields in Canada found a significantly greater abundance of wild bees in organic fields (averages 86 bees per organic field sample, 58 bees per conventional field; Morandin & Winston 2005).

A comparison of 21 organic and 21 conventional winter wheat fields in northern Germany found a greater abundance and diversity of wild bees on organic fields than on paired control fields (Kleijn *et al.* 2006, Holzschuh *et al.* 2007). Average bee species richness per field was 6.9 for organic fields and 2.1 species for conventional fields. 1,326 individuals of 31 bee species (average abundance 63.1) were recorded in organic fields compared to 181 individuals of 16 species (average abundance 8.6) in conventional fields.

In the same study, the total number of bee species was higher under organic farming whether you considered the number found at individual sites, the total number found in each region or the total for the entire study (Clough *et al.* 2007). Diversity between sites as well as within sites was greater for organic fields than for conventional fields. This means bee diversity improved under organic wheat farming at the larger landscape level, as well as the local level.

Also in the same study, Holzschuh *et al.* (2008) report higher bee abundance and diversity on permanent fallow strips next to organic winter fields, compared to fallow strips next to conventional wheat fields. On average, 2.6 m wide annually mown fallow strips next to organic fields had 6.3 bee species, 8.5 bumblebee individuals and 2.6 solitary bees/100 m in total over four surveys, compared to 4.0 species, 3.7 bumblebees and 1.1 solitary bees/100 m on strips next to conventional fields.

A study of 15 organic and 40 conventional arable field boundaries in Finland found no significant difference in the numbers of bumblebees or bumblebee species (Ekroos *et al.* 2008). On average, three bumblebees from 1.1 species were recorded per transect on conventional farm field boundaries, and 3.8 bumblebees from 1.4 species on organic farm field boundaries.

Rundlöf *et al.* (2008) surveyed bumblebees *Bombus* spp. on 12 pairs of organic and conventional farms in Sweden, and found significantly more bumblebees and bumblebee species on organic than conventional farms (on average 7.7 and 4.9 species/farm on organic and conventional farms respectively). This difference between organic and conventional farms was not statistically significant when only the six pairs of farms in heterogenous (mixed) farming landscapes, with smaller field sizes and more grassland, were considered. So organic farming had a greater effect on wild bumblebees in intensive, homogenous arable landscapes.

Winfree *et al.* (2008) surveyed wild solitary and social bees visiting flowering crops on 22 or 23 farms, of which six or seven were organic and 16 conventional, in Pennsylvania and New Jersey, USA. Organic and conventional farms did not differ in field size, crop diversity or wild/weedy plant diversity and all lay in a heterogeneous landscape with many small patches of natural habitat such as woodland. They found no difference in either the abundance or species richness of bees between organic and conventional farms.

Belfrage K., Björklund J. & Salomonsson L. (2005) The effects of farm size and organic farming on diversity of birds, pollinators and plants in a Swedish landscape. *Ambio*, 34, 582–588. www↗

Clough Y., Holzschuh A., Gabriel D., Purtauf T., Kleijn D., Kruess A., Steffan-Dewenter I. & Tscharntke T. (2007) Alpha and beta diversity of arthropods and plants in organically and conventionally managed wheat fields. *Journal of Applied Ecology*, 44, 804–812. www↗

Ekroos J., Piha M. & Tiainen J. (2008) Role of organic and conventional arable field boundaries on boreal bumblebees and butterflies. *Agriculture, Ecosystems and Environment*, 124, 155–159. www↗

Holzschuh A., Steffan-Dewenter I., Kleijn D. & Tscharntke T. (2007) Diversity of flower-visiting bees in cereal fields: effects of farming system, landscape composition and regional context. *Journal of Applied Ecology*, 44, 41–49. www↗

Holzschuh A., Steffan-Dewenter I. & Tscharntke T. (2008) Agricultural landscapes with organic crops support higher pollinator diversity. *Oikos*, 117, 354–361. www↗

Kleijn D., Baquero R.A., Clough Y., Diaz M., De Esteban J., Fernandez F., Gabriel D., Herzog F., Holzschuh A., Johl R., Knop E., Kruess A., Marshall E.J.P., Steffan-Dewenter I., Tscharntke T., Verhulst J., West T.M. & Yela J.L. (2006) Mixed biodiversity benefits of agri-environment schemes in five European countries. *Ecology Letters*, 9, 243–254. www↗

Morandin L.A., & Winston M.L. (2005) Wild bee abundance and seed production in conventional, organic, and genetically modified canola. *Ecological Applications*, 15, 871–881. www↗

Rundlöf M., Nilsson H. & Smith H.G. (2008) Role of organic and conventional field boundaries on boreal bumblebees and butterflies. *Biological Conservation*, 141, 417–426. www↗

Winfree R., Williams N., Gaines H., Ascher J.S. & Kremen C. (2008) Wild bee pollinators provide majority of crop visitation across land-use gradients in New Jersey and Pennsylvania, USA. *Journal of Applied Ecology*, 45, 793–802. www↗

4.3 Restrict certain pesticides

- One site comparison study in Italy showed that a reduction in the number of solitary bee species in late summer associated with repeated applications of the insecticide fenitrothion can be avoided by not applying the insecticide.

Brittain *et al.* (2010) compared the wild bee and butterfly communities in 17 conventional grapevine fields with those in four vine fields in a natural park with negligible insecticide use, in Veneto, northeastern Italy. Sites with and without insecticide treatments had different landscape features and sample sizes in this study, so direct comparison is difficult. However, the study found that a reduction in the number of wild bee species caught in pan traps in July and August, apparently associated with two or more applications of the insecticide fenitrothion, did not happen in vine fields that were not treated. Bumblebees, counted in transect walks, were not affected by fenitrothion applications in this way.

We have not found any evidence of the effects on wild bees of restricting neonicotinoid pesticides such as imidacloprid, although their use on some flowering crops has recently been suspended or banned in France, Germany, Italy and Slovenia to protect honey bees (Kindemba 2009).

Brittain C.A., Vighi M., Bommarco R., Settele J. & Potts S.G. (2010) Impacts of a pesticide on pollinator species richness at different spatial scales. *Basic and Applied Ecology*, 11, 106–115. www↗

Kindemba V. (2009) *The impact of neonicotinoid insecticides on bumblebees, honey bees and other non-target invertebrates.* Buglife Report, ISBN 978-1-904878-964. Available at: http://www.buglife.org.uk/Resources/Buglife/revised%20neonics%20 report.pdf

4.4 Reduce pesticide or herbicide use generally

See also Chapter 4.1 *Introduce agri-environment schemes that reduce spraying.*

• One replicated trial in the USA showed that numbers of foraging bees on squash farms are not affected by the responsible use of pesticides.

Shuler *et al.* (2005) compared the abundance of bees visiting squash flowers *Cucurbita* sp. on farms that either used pesticides (13 farms) or did not (12 farms), in the eastern USA. They found no difference in the abundances of squash bees *Peponapis pruinosa*, bumblebees *Bombus* sp. or honey bees *Apis mellifera* that could be explained by pesticide use. The study included no information about the type of pesticide, quantity or timing of its use. The authors assumed pesticides were applied on these study farms at times when bees were not exposed.

A large replicated trial of the effects of farmland management on biodiversity in the UK found that switching to the broad spectrum herbicides used with herbicide tolerant genetically modified crops reduced bee abundance in oilseed rape *Brassica napus* ssp. *oleifera* and beet *Beta vulgaris* ssp. *vulgaris* fields, but not in maize *Zea mays* fields or field margins (Hawes *et al.* 2003, Roy *et al.* 2003, Bohan *et al.* 2005). Whilst these results demonstrated the potential impact of changing the herbicide regime on wild bees, they are not included on Conservation Evidence, because neither the intervention (switch to broad spectrum herbicide) nor its avoidance (conventional herbicide treatment) could be construed as an intervention intended to conserve wildlife.

Bohan D.A., Boffey C.W.H., Brooks D.R., Clark S.J., Dewar A.M., Firbank L.G., Haughton A.J., Hawes C., Heard M.S., May M.J., Osborne J.L., Perry J.N., Rothery P., Roy D.B., Scott R.J., Squire G.R., Woiwod I.P. & Champion G.T. (2005) Effects on weed and invertebrate abundance and diversity of herbicide management in genetically modified herbicide-tolerant winter-sown oilseed rape. *Proceedings of the Royal Society B*, 272, 463–474.

Hawes C., Haughton A., Osborne J.L., Roy D., Clark S., Perry J., Rothery P., Bohan D., Brooks D., Champion G., Dewar A., Heard M., Woiwod I., Daniels R., Young M., Parish A., Scott R., Firbank L. & Squire G. (2003) Responses of plants and invertebrate trophic groups to contrasting herbicide regimes in the Farm Scale Evaluations of genetically modified herbicide-tolerant crops. *Philosophical Transactions of the Royal Society B*, 358, 1899–1913.

Roy D.B., Bohan D.A., Haughton A.J., Hill M.O., Osborne J.L., Clark S.J., Perry J.N., Rothery P., Scott R.J., Brooks D.R., Champion G.T., Hawes C., Heard M.S. & Firbank L.G. (2003) Invertebrates and vegetation of field margins adjacent to crops subject to contrasting herbicide regimes in the Farm Scale Evaluations of genetically modified herbicide-tolerant crops. *Philosophical Transactions of the Royal Society of London Series B*, 358, 1879–1898.

Shuler R.E., Roulston T.H. & Farris G.E. (2005) Farming practices influence wild pollinator populations on squash and pumpkin. *Journal of Economic Entomology*, 98, 790–795. www⬈

4.5 Reduce fertilizer run-off into margins

See also Chapter 3.10 *Provide grass strips at field margins.*

• We have captured no evidence on the effects of specific interventions to reduce fertilizer run off into field margins.

4.6 Leave field margins unsprayed within the crop (conservation headlands)

- Two replicated controlled trials in England showed that conservation headlands do not attract more foraging bumblebees than conventional crop fields. One replicated trial found fewer bees on conservation headlands than in naturally regenerated, uncropped field margins in England.

Background

Conservation headland management involves restricted herbicide and insecticide spraying in a 6 m margin of sown arable crop. The prescription allows selected herbicide applications to control pernicious weeds or invasive alien species.

Kells *et al.* (2001) counted bumblebees *Bombus* spp. and honey bees *Apis mellifera* on 50 m transects in five 6 m wide field margins managed as conservation headlands, and ten naturally regenerated, uncropped field margins, in the West Midlands, UK. They recorded averages of less than three bees/transect in conservation headlands, compared to averages of between 10 and 50 bees/transect in naturally regenerated margins.

A replicated controlled trial (Pywell *et al.* 2005) in East Anglia and the West Midlands, UK, found no significant difference in bumblebee species richness and abundance when 16 conservation headlands were compared with paired conventional field margins. In both types of field margin, a few species of plant contributed to the vast majority of foraging visits by bumblebees, mainly creeping thistle *Cirsium arvense* and spear thistle *C. vulgare*.

In a replicated controlled trial at six sites (two replicates/site) across central and eastern England, Carvell *et al.* (2007) found that unsprayed conservation headlands did not support more bumblebee individuals or species than conventional cropped field margins.

Carvell C., Meek W.R., Pywell R.F., Goulson D. & Nowakowski M. (2007) Comparing the efficacy of agri-environment schemes to enhance bumble bee abundance and diversity on arable field margins. *Journal of Applied Ecology*, 44, 29–40. www↗

Kells A.R., Holland J. & Goulson D. (2001) The value of uncropped field margins for foraging bumblebees. *Journal of Insect Conservation*, 5, 283–291. www↗

Pywell R.F., Warman E.A., Carvell C., Sparks T.H., Dicks L.V., Bennett D., Wright A., Critchley C.N.R. & Sherwood A. (2005) Providing foraging resources for bumblebees in intensively farmed landscapes. *Biological Conservation*, 121, 479–494. www↗

5. Threat: transportation and service corridors

Key Messages

Restore species-rich grassland on road verges

One replicated controlled trial showed that road verges planted with native prairie vegetation in Kansas, USA supported a greater number and diversity of bees than frequently mown grassed verges.

Manage land under power lines for wildlife

One replicated trial in Maryland, USA found more bee species under power lines managed as scrub than in equivalent areas of annually mown grassland.

5.1 Restore species-rich grassland on road verges

- One replicated controlled trial showed that road verges planted with native prairie vegetation in Kansas, USA supported a greater number and diversity of bees than frequently mown grassed verges.

A replicated controlled trial in Kansas, USA (Hopwood 2008) found that seven road verges planted with native prairie grasses and flowers supported a greater number and diversity of bees than paired conventionally managed verges, four to five years after planting. Restored verges were mown every two to four years, or burned annually, while conventionally managed verges were mown three to four times during each growing season and certain weeds treated with herbicide. In total, 812 bees from 79 species were found on restored verges, compared to 353 bees from 53 species on conventionally managed verges. The verges studied were all 18–84 m wide. Verge width, slope, aspect and density of traffic on the adjacent road made no difference to the bee community. Native prairie vegetation includes bunch grasses, which grow in a way that leaves bare ground exposed and provides potential nesting areas for ground-nesting bees.

Hopwood J.L. (2008) The contribution of roadside grassland restorations to native bee conservation. *Biological Conservation*, 141, 2632–2640. www↗

5.2 Manage land under power lines for wildlife

• One replicated trial in Maryland, USA found more bee species under power lines managed as scrub than in equivalent areas of annually mown grassland.

Power line rights-of-way are unfarmed and provide potentially valuable linear strips of habitat for bees and other wildlife. In the USA, they are periodically cleared of vegetation by mowing and/or herbicide treatment. A more cost-effective management method involves removing trees and other tall vegetation, mechanically and with selective herbicides, but retaining a dense scrub. One replicated trial under eight power line strips on a Wildlife Refuge in Maryland, USA (Russell *et al.* 2005) found significantly more bee species under power lines managed this way (32.5 bee species/site on average) than in equivalent areas of annually mown grassland on the Refuge, representing conventional power line management (23.2 species/site). There was no significant difference between power line scrub and mown grassland in the abundance of bees.

Russell K.N., Ikerd H. & Droege S. (2005). The potential conservation value of unmowed powerline strips for native bees. *Biological Conservation*, 124, 133–148. www↗

6. Threat: biological resource use

Key Messages

Sustainable management of wild honey bees

We have found no direct evidence of the impact of reduced honey-hunting or improved harvesting methods on wild bee populations.

Replace honey-hunting with apiculture

One study reported that a programme to enhance take-up of stingless beekeeping in southern Mexico increased the number of managed colonies in the area. Five trials in Central or South America contributed to the scientific improvement of stingless beekeeping methods.

Legally protect large native trees from logging

A study in Brazil showed that the species *Melipona quadrifasciata* selectively nested in the protected cerrado tree *Caryocar brasiliense*, suggesting that protecting this species from logging or wood harvesting has helped to conserve stingless bees.

Re-plant native forest

We have captured no evidence on the effects of reforestation on wild bee communities or populations.

Retain dead wood in forest management

We have captured no evidence on the effects of retaining dead wood on wild bee communities or populations in woodland or forest.

Hunting and collecting terrestrial animals

Background

In Asia, Africa, South America, Australia and parts of Europe, native honeybees of the genus *Apis* or the family Meliponinae (stingless bees) have been traditionally managed domestically for their honey, or honey is harvested from the wild.

The diversity of these native species is threatened in many tropical and subtropical areas. The threat is partly from land use changes such as deforestation, but it is accompanied by a decline in traditional beekeeping practices. In these circumstances beekeeping itself can represent a conservation measure to help sustain populations of some species, whilst sustaining local livelihoods and improving people's understanding of the value of natural habitat.

6.1 Manage wild honey bees sustainably

- We can find no evidence of the impact of reduced honey-hunting or improved harvesting methods on wild bee populations. One trial in southern Vietnam, showed that occupancy of artificial rafters by the giant honey bee *Apis dorsata* can be over 85% when rafters are placed by a large clearing greater than 25 m in diameter.

Background

Honey harvesting or honey-hunting from wild *Apis* colonies is a common and traditional practice in parts of Asia, one that is considered to pose a potentially serious threat to populations of some wild bee species (Oldroyd & Nanork 2009).

In southern Vietnam, a form of beekeeping exists in which honey is harvested repeatedly from wild colonies of the giant honey bee *Apis dorsata* without destroying the combs, by persuading the bees to form colonies on easily accessible artificial rafters. Rafters are split tree trunks, erected on poles at an angle of 15–35° to the horizontal. A trial of 507 rafters erected by beekeepers in U Minh Forest, Minh Hai Province (Tan *et al.* 1997), showed that occupancy by bees was significantly higher when the open space in front of the rafter was very large, over 25 m in diameter (85% and 92% of rafters occupied in dry and rainy seasons respectively, compared to 33–51% for open spaces from 3 to 25 m in diameter).

We can find no direct evidence of the effects of reduced honey hunting or improved honey-harvesting methods on wild bee populations.

Oldroyd B.P. & Nanork P. (2009) Conservation of Asian honeybees. *Apidologie*, 40, 296–312.

Tan N.Q., Chinh P.H., Thai P.H. & Mulder V. (1997) Rafter beekeeping with *Apis dorsata*: some factors affecting the occupation of rafters by bees. *Journal of Apicultural Research*, 36, 49–54. www↗

6.2 Replace honey-hunting with apiculture

- One study reported that a programme to enhance take-up of stingless beekeeping in southern Mexico increased the number of managed colonies in the area.

- Five trials contributed to scientific improvement of stingless beekeeping methods. Two controlled trials showed that either brewer's yeast (one trial) or a mix with 25% pollen collected by honey bees *Apis mellifera* (one trial) can be used as a pollen substitute to feed *Scaptotrigona postica* in times of pollen scarcity. A study on the island of Tobago found a wooden hive design with separate, different-shaped honey and brood chambers allowed honey to be extracted without damaging the brood. One trial showed that 50 g of comb with mature pupae is enough to start a new daughter colony of *S. mexicana*. One trial found brood growth was higher in traditional log hives than in box hives with internal volumes exceeding 14 litres, and recommended smaller box hives.

- We have captured no clear evidence about whether these activities help conserve bees or enhance native bee populations.

Background

Traditionally, stingless bees have been kept in hollow logs in Central and South America, but these make honey extraction and parasite control difficult and so improved methods are being developed.

Two controlled trials in Ribeirão Prêto, São Paulo, Brazil tested different pollen substitute diets for their ability to support development in stingless bee workers of the species *Scaptotrigona postica*. One trial with two groups of 10 bees for each diet found brewer's yeast was the best pollen substitute, leading to better development of the ovaries and hypopharyngeal gland than two brands of commercially available pollen substitute, or bulrush *Typha* pollen. The control group, fed on pollen collected by other *S. postica* bees, developed better than all the other groups (Zucoloto 1977).

The second trial, with groups of 15 worker bees given each experimental diet found that a mix of *S. postica*-collected pollen with 25% *Apis mellifera*-collected pollen allowed equivalent development in *S. postica* workers to pure *S. postica* pollen, but higher proportions of *A. mellifera* pollen in the mix led to reduced development and lower pollen consumption (Testa *et al.* 1981).

Sommeijer (1999) described a hive design, the 'Utrecht University-Tobago Hive' (UTOB hive), with separate and different-shaped honey and brood chambers. Three years of testing on the island of Tobago found that the stingless bee species *Melipona favosa* formed colonies in this type of hive with brood confined to the brood chamber and a single layer of honey and pollen pots in the honey chamber. Honey could then be extracted with little disturbance to the brood or pollen stores.

A replicated trial with wild-caught colonies of the stingless bee *Scaptotrigona mexicana*, endemic to Mexico and Guatemala, demonstrated that new colonies can be propagated from old colonies with 50 g of brood, containing approximately 2,750 mature pupae, along with 3,000-4,000 workers, 100 g of honey and 10 g of wax (Arzaluz *et al.* 2002). Five colonies started with 50 g of brood and five started with 90 g of brood did not differ in their average weight gain over 10 weeks.

In a replicated trial, Quezada-Euan & González-Acereto (1994) found that brood growth was faster in colonies of *Melipona beecheii* housed in traditional log hives (internal volume 10 litres) than in those housed in more modern box hives (internal volumes 14.3 and 14.5 litres). The authors suggested this is due to difficulties with the bees' ability to regulate temperature. They recommended reducing the internal volume of box hives by about one third.

González-Acereto *et al.* (2006) report results of a programme of measures to promote beekeeping with native stingless bees in the state of Yucatán, Mexico. The program involved setting up a central bank of colonies available on loan, providing training courses and support for beekeepers and developing beekeeping techniques, new uses for stingless bees and their products. Around 150 people were trained in stingless beekeeping over five years, and this resulted in 324 new colonies being kept. After six years, the colony bank, developed with colonies obtained from the wild after clearance of forest patches, contained 377 colonies of 10 native stingless bee species.

Arzaluz A., Obregón F. & Jones R. (2002) Optimum brood size for artificial propagation of the stingless bee *Scaptotrigona mexicana*. *Journal of Apicultural Research*, 41, 62–63. www↗

González-Acereto J.A., Quezada_Euán J.J.G. & Medina-Medina L.A. (2006). New perspectives for stingless beekeeping in the Yucatan: results of an integral program to rescue and promote the activity. *Journal of Apicultural Research*, 45, 234–239. www↗

Quezada-Euan J.J.G. & Gonzalez-Acereto J. (1994) A preliminary study on the development of colonies of *Melipona beecheii* in traditional and rational hives. *Journal of Apicultural Research*, 33, 167–170. www↗

Testa P.R., Silva A.N. & Zucoloto F.S. (1981) Nutritional value of different pollen mixtures for *Nannotrigona (Scaptotrigona) postica*. *Journal of Apicultural Research*, 20, 94–96. www↗

Sommeijer M.J. (1999) Beekeeping with stingless bees: a new type of hive. *Bee World*, 80, 70–79. www↗

Zucoloto F.S. (1977) Nutritive value of some pollen substitutes for *Nannotrigona (Scaptotrigona) postica*. *Journal of Apicultural Research*, 16, 59–61. www↗

Logging and wood harvesting

6.3 Legally protect large native tress

- A study in degraded savannah in Minas Gerais, Brazil showed that the stingless bee species *Melipona quadrifasciata* selectively nested in the protected cerrado tree *Caryocar brasiliense*, evidence that protecting this species from logging or wood harvesting has helped to conserve stingless bees.

Background

There is evidence that social bees such as honey bees and stingless bees prefer to nest in trees above a certain size, or girth, and in isolated trees (for example Eltz *et al.* 2003, Thomas *et al.* 2009). For this reason, specific protection of individual large trees in habitats undergoing degradation has the potential to help sustain bee populations.

The cerrado tree *Caryocar brasiliense* is the only tree species protected by federal regulations in Brazil. A detailed study of nest sites used by the stingless bee species *Melipona quadrifasciata* in 18 km² of degraded cerrado (72 plots, each 500 m²) in Minas Gerais, Brazil, found that they almost exclusively nested in *C. brasiliense* (Antonini & Martins 2003). Forty-six out of 48 nests were found in that species, although there were 55 tree species at the site. The authors argue that *M. quadrifasciata* is only found in the area because of the protection of *C. brasiliense*.

Antonini Y. & Martins R.P. (2003) The value of a tree species (*Caryocar brasiliense*) for a stingless bee *Melipona quadrifasciata quadrifasciata*. *Journal of Insect Conservation*, 7, 167–174. www↗

Eltz T., Bruhl C.A., Imiyabir Z. & Linsenmair K.E. (2003) Nesting and nest trees of stingless bees (Apidae: Meliponini) in lowland dipterocarp forests in Sabah, Malaysia, with implications for forest management. *Forest Ecology and Management*, 172, 301–313.

Thomas S.G., Varghese A., Roy P., Bradbear N., Potts S.G. & Davidar P. (2009) Characteristics of trees used as nest sites by *Apis dorsata* (Hymenoptera, Apidae) in the Nilgiri Biosphere Reserve, India. *Journal of Tropical Ecology*, 25, 559–562.

6.4 Re-plant native forest

• We have found no evidence on the impact of reforestation on wild bee communities or populations.

6.5 Retain dead wood in forest management

See Chapter 10 *Provide artificial nest sites for solitary bees,* for one trial in which nest boxes were placed in dead standing trees in lowland tropical rainforest (Thiele 2005).

- We have found no evidence on the impact of retaining dead wood in forests or woodlands on wild bee communities or populations.

Thiele R. (2005) Phenology and nest site preferences of wood-nesting bees in a Neo-tropical lowland rain forest. *Studies on Neotropical Fauna and Environment*, 40, 39–48. www↗

7. Threat: natural system modification - natural fire and fire suppression

Key Messages

Control fire risk using mechanical shrub control and/or prescribed burning.

One trial in the USA showed that for bee conservation, it is best to control fire using cutting and burning combined.

7.1 Control fire risk using mechanical shrub control and/or prescribed burning

- One replicated controlled trial in mixed temperate forest in the USA showed that for bee conservation, it is best to control fire using cutting and burning combined. This increases herbaceous plant cover in subsequent years.

Background

Natural fire has been shown to have an initially adverse effect on the abundance of wild solitary bees, followed by rapid recovery, in dry Mediterranean shrubland in Israel (Ne'eman *et al.* 2000, Potts *et al.* 2003).

However, we have not found evidence of the effects of using or control-ling fire as a direct management strategy for conservation purposes in this habitat.

For butterflies, there is evidence that leaving permanently unburnt fire refuges is beneficial (for example, Swengel & Swengel 2007), which may also be true for bees. We have found no evidence of the effects of this strategy on bees.

A replicated controlled trial in mixed temperate forests in North Caro-lina, USA, tested the effects of prescribed burning and mechanical shrub control (cutting) or both, on the abundance of flower-visiting insects in the subsequent two years (Campbell *et al.* 2007). There were three replicates of each treatment, in 14 ha plots. There were more bees and more bee species in plots that underwent both mechanical shrub control and prescribed burn-ing, compared to plots with mechanical shrub control only, prescribed burn only or no fire control. Mechanical shrub control and burning combined led to hotter fires and increased herbaceous plant cover, providing more forage plants for bees in subsequent years.

Campbell, J.W., Hanula, J.L. & Waldrop, T.A. (2007) Effects of prescribed fire and fire surro-gates on floral visiting insects of the Blue Ridge province in North Carolina. *Biologi-cal Conservation*, 134, 393–404. www↗

Ne'eman G., Dafni A. & Potts S.G. (2000) The effect of fire on flower visitation rate and fruit set in four core-species in east Mediterranean scrubland. *Plant Ecology*, 146, 97–104.

Potts S.G., Vulliamy B., Dafni A., Ne'eman G., O'Toole C., Roberts S. & Willmer P.G. (2003) Response of plant-pollinator communities following fire: changes in diversity, abun-dance and reward structure. *Oikos*, 101, 103–112.

Swengel A.B. & Swengel S.R. (2007) Benefit of permanent non-fire refugia for Lepidoptera conservation in fire-managed sites. *Journal of Insect Conservation*, 11, 263–279.

8. Threat: invasive non-native species

Eradicate existing populations

One replicated trial in the USA demonstrated that invasive Africanized honey bee colonies *Apis mellifera* can be killed using insecticide in syrup bait. One replicated controlled before-and-after trial attempted to eradicate the European buff-tailed bumblebees *Bombus terrestris* from trial sites in Japan by catching and killing foraging bees. The treatment failed.

Control deployment of hives/nests

We have found no direct evidence of the effects of excluding *Apis mellifera* hives, or nests of other managed pollinators, on populations of wild bees.

Prevent escape of commercial bumblebees from greenhouses

Two trials have tested methods to keep bumblebees within greenhouses. One trial in Canada showed that a greenhouse covering that transmits ultraviolet light reduced the number of bees escaping. One trial in Japan showed that externally mounted nets and zipped, netted entrances can keep commercial bumblebees inside greenhouses.

Prevent introduction and spread of the small hive beetle

One replicated trial in the USA tested the effect of mite-killing strips in commercial honey bee *Apis mellifera* transport packages. More than half the beetles escaped the packages and were not killed by the strip.

Ensure commercial hives/nests are disease free

One randomised controlled trial in Canada found that the antibiotic fumagillin is not effective against *Nosema bombi* infection in managed colonies of the western bumblebee *Bombus occidentalis*. One replicated controlled trial in South Korea found that Indian meal moth *Plodia interpunctella* in commercial bumblebee colonies can be controlled with the insect pathogen *Bacillus thuringiensis*.

Keep pure breeding populations of native honey bee subspecies

One replicated trial in Switzerland found that 'pure breeding' populations of the European black honey bee *Apis mellifera mellifera* contained a significant proportion (28%) of hybrids with an introduced subspecies *Apis mellifera carnica*.

Exclude introduced European earwigs from nest sites

In California, USA, a replicated controlled trial showed that numbers of introduced European earwigs *Forficula auricularia* resting in solitary bee nest boxes were reduced using a sticky barrier ('Tanglefoot'), increasing use of the boxes by native bees.

Background

Several non-native bee species present a risk as invasives in various parts of the world at present, through direct competition and the introduction of non-native parasites and pathogens.

The managed European or Africanized honey bee *Apis mellifera* is widely introduced and naturalized. Non-native subspecies of *A. mellifera* are managed in many countries in Europe where other subspecies are, or were once, native (De la Rúa *et al.* 2009). Species of European leafcutter bee *Megachile* spp. managed for pollination have become naturalised in the western USA.

Since the start of commercial bumblebee rearing in 1987, non-native bumblebees have been introduced to more than 11 countries in the Americas, Asia and Australasia. The European buff-tailed bumblebee *Bombus terrestris*, colonies of which have been widely imported for pollination of greenhouse crops, has become naturalised in Tasmania and Japan. There are no native bumblebees in Tasmania, but in Japan *B. terrestris* has been shown to have negative impacts on native bumblebee species, apparently through competition for nest sites (Inoue *et al.* 2008). At least four non-native species of *Bombus* are established in Argentina, and non-native *B. terrestris* have been recently been found to be carrying two internal parasites, *Crithidia bombi* and *Apicystis bombi*, not widely found in native Argentinian *Bombus* species (Plischuk & Lange 2009).

Non-native subspecies of *B. terrestris* are still being introduced in places where other subspecies are native, such as the UK, posing a potential competitive or hybridisation threat (for example Ings *et al.* 2006).

Bumblebee colonies kept in greenhouses for commercial pollination can hold higher levels of parasite infection than wild colonies. A study in Canada showed that native bumblebees nesting close to greenhouses were infected with one such parasite, the protozoan *Crithidia bombi*, but native colonies elsewhere in the same region were free of this parasite (Colla *et al.* 2006). The observed decline in five North American bumblebee species has been blamed at least partly on the microsporidian *Nosema bombi* and other pathogens introduced in commercial bumblebee colonies (Otterstatter & Thomson 2008, Stout & Morales 2009).

Invasive alien plants also interact with wild bees, either directly by providing forage to certain species, or indirectly by altering native plant communities (Stout & Morales 2009). However, the impact of invasive plants on bee communities is poorly researched and understood. At present we know of no examples where the control of alien invasive plants is advised as a bee conservation strategy.

Colla S.R., Otterstatter M.C., Gegear R.J. & Thomson J.D. (2006) Plight of the bumble bee: Pathogen spillover from commercial to wild populations. *Biological Conservation*, 129, 461–467.

De la Rúa P., Jaffé R., Dall'Olio R., Muñoz I. & Serrano J. (2009) Biodiversity, conservation and current threats to European honeybees. *Apidologie*, 40, 263–284.

Ings T.C., Ward N.L. & Chittka L. (2006) Can commercially imported bumble bees out-compete their native conspecifics? *Journal of Applied Ecology*, 43, 940–948.

Inoue M.N., Yokoyama J. & Washitani I. (2008) Displacement of Japanese native bumble-bees by the recently introduced *Bombus terrestris* (L.) (Hymenoptera: Apidae). *Journal of Insect Conservation*, 12, 135–146.

Otterstatter M.C. & Thomson J.D. (2008) Does pathogen spillover from commercially reared bumble bees threaten wild pollinators? *Plos One*, 3, article e2771.

Plischuk S. & Lange C.E. (2009) Invasive *Bombus terrestris* (Hymenoptera: Apidae) parasitized by a flagellate (Euglenozoa: *Kinetoplastea)* and a neogregarine (Apicomplexa: Neogregarinorida). *Journal of Invertebrate Pathology*, 102, 261–263.

Stout J.C. & Morales C.L. (2009) Ecological impacts of invasive alien species on bees. *Apidologie*, 40, 388–409.

8.1 Eradicate existing populations

• One replicated trial in Louisiana, USA, demonstrated that colonies of invasive Africanized honey bees *Apis mellifera* can be killed by providing insecticide (acephate)-laced syrup for 30 minutes.

• One replicated controlled before-and-after trial attempted to eradicate European buff-tailed bumblebees *Bombus terrestris* from trial sites in Japan by catching and killing foraging bees. The treatment led to an increase in numbers of two native bumblebee species, but did not eradicate *B. terrestris*.

Williams *et al.* (1989) tested a method for killing naturalised colonies of Africanized honey bee *Apis mellifera* using poisoned bait, in an outdoor experiment in Louisiana, USA. Nineteen colonies were given sucrose-honey syrup containing the organophosphate insecticide acephate at 250 ppm (mg/l), at feeding stations 10 m away from the experimental hives during April 1988. 13 colonies died within three days. In six treated colonies and two control colonies, the queen bee survived or was replaced and the colony survived. Treatment lasted for 30–40 minutes before foragers became poisoned. No other insects were observed visiting the bait.

Nagamitsu *et al.* (2010) removed foraging non-native buff-tailed bumblebees *B. terrestris* from six wooded sites (0.1–1.0 ha in size) in agricultural and urban landscapes in the Chitose River basin, Ishikari, Hokkaidō, Japan, from 2005 to 2006, and monitored bumblebees at these and seven control sites from 2004 to 2006. The removal treatment increased numbers of the native bumblebee *B. ardens* in both years, and increased numbers of *B. hypocrita* in one year, but did not consistently lead to a drop in the numbers of *B. terrestris* trapped at the sites.

Nagamitsu T., Yamagishi H., Kenta T., Inari N. & Kato E. (2010) Competitive effects of the exotic *Bombus terrestris* on native bumble bees revealed by a field removal experiment. *Population Ecology*, 52, 123–136. www↗

Williams J.L., Danka R.G. & Rinderer T.E. (1989) Baiting system for selective abatement of undesirable honey bees. *Apidologie*, 20, 175–179. www↗

8.2 Control deployment of hives/nests

• We have found no direct evidence of the effects of excluding *Apis mellifera* hives, or nests of other managed pollinators, on populations of wild bees.

Background

Being near honey bee *Apis mellifera* hives has been shown to reduce worker size (Goulson & Sparrow 2009), forager return rates, the proportion of foragers collecting pollen and the number of reproductives produced (Thomson 2004), for bumblebee species in the UK and the USA. However, we know of no direct evidence of a positive effect of excluding *A. mellifera* hives on populations of bumblebees.

Goulson D. & Sparrow K.R. (2009) Evidence for competition between honey bees and bumblebees; effects on bumblebee worker size. *Journal of Insect Conservation*, 13, 151-163.

Thomson D. (2004) Competitive interactions between the invasive European honey bee and native bumble bees. *Ecology*, 85, 458–470.

8.3 Prevent escape of commercial bumblebees from greenhouses

• One small replicated trial in Canada showed that a plastic greenhouse covering that transmits ultraviolet light (so transmitted light is similar to daylight) reduced the numbers of bumblebees from managed colonies escaping through open gutter vents. One trial in Japan showed that externally mounted nets and zipped, netted entrances can keep commercial bumblebees inside greenhouses as long as they are regularly checked and maintained.

We have captured two studies about the efficacy of efforts to confine commercial bees within greenhouses.

A small replicated trial in Ontario, Canada, (Morandin *et al.* 2001) showed that loss of bees from commercially managed colonies of the common eastern bumblebee *Bombus impatiens* in greenhouses was much lower under a type of plastic covering that transmitted ultraviolet light (wavelengths 300–350 nm) than under coverings that blocked this kind of light. Counts were taken in greenhouses in March, when outside temperatures are too low for bumblebees to survive. After 10 day observation periods in three greenhouses of each type of covering, colonies under the plastic transmitting UV had an average of 86 bees per colony remaining, while colonies under other types of plastic covering had an average of 36 bees per colony. The authors suggest bees escaped through open gutter vents, which they cannot see so easily when there is less contrast (in the ultraviolet part of the spectrum) between daylight and light coming through the greenhouse roof.

Koide *et al.* (2008) tested whether netting could prevent the escape of the buff-tailed bumblebee *B. terrestris* from four greenhouses with different netting techniques in Japan, where netting is a legal requirement for greenhouse growers using bumblebee colonies. The study showed that nets mounted on the outside of windows with packers (tubes that hold plastic film) or Vinipets (U-shaped devices) prevented bumblebee escape, providing the nets were regularly checked and maintained. Nets mounted on the inside, or on the outside with clips, allowed bees to escape. Double netting of doors, even with a plastic vestibule, also allowed bumblebees to escape, but zipped, netted entrances prevented escape as long as the entrance was weighted at the bottom.

Koide T., Yamada Y., Yabe K. & Yamashita F. (2008) Methods of netting greenhouses to prevent the escape of bumblebees. *Japanese Journal of Applied Entomology and Zoology*, 52, 19–26. www↗

Morandin L.A., Laverty T.M., Kevan P.G., Khosla S. & Shipp L. (2001) Bumble bee (Hymenoptera: Apidae) activity and loss in commercial tomato greenhouses. *The Canadian Entomologist*, 133, 883–893. www↗

8.4 Prevent spread of the small hive beetle

• One replicated trial in the USA tested the effect of using mite-killing strips in commercial honey bee *Apis mellifera* transport packages, to reduce the spread of small hive beetle. More than half the beetles escaped the packages and were not killed by the strip.

Background

The small hive beetle *Aethina tumida*, a native of sub-Saharan Africa, is invading in North America, Australia and southern Europe and is perceived to pose a particular threat to bumblebees (Neumann & Ellis 2008). One experimental study shows that this species can substantially reduce bumblebee worker numbers in artificial colonies (Ambrose *et al.* 2000).

There is a substantial literature on methods to control small hive beetle within honey bee colonies, but we do not consider these studies to represent conservation interventions. We have captured one experimental study of the effects of efforts to reduce the spread of this species.

In a replicated, controlled trial, Baxter *et al.* (1999) tested methods to control the spread of small hive beetle in packages for transporting honey bees *Apis mellifera* for the beekeeping industry in Texas, USA. They placed 'Checkmite strips' (containing the organophosphate coumaphos) in various positions inside or on the packages and intentionally introduced ten or twenty adult beetles. More than half the beetles escaped from the packages through a ventilation panel and were not trapped or recovered. A Checkmite strip hanging in the middle of the package killed 94% of the remaining beetles. Beetles were not lured out of the packages by light traps.

Ambrose J.T., Stanghellini M.S. & Hopkins D.I. (2000) A scientific note on the threat of small hive beetles (*Aethina tumida* Murray) to bumblebee (*Bombus* spp.) colonies in the United States. *Apidologie*, 31, 455–456.

Baxter J.R., Elzen P.J., Westervelt D., Causey D., Randall C., Eischens F.A. & Wilson W.T. (1999) Control of the small hive beetle, *Aethina tumida* in package bees. *American Bee Journal*, 139, 792–793. www↗

Neumann P. & Ellis J.D. (2008) The small hive beetle (*Aethina tumida* Murray, Coleoptera: Nitidulidae): distribution, biology and control of an invasive species. *Journal of Apicultural Research and Bee World*, 47, 181–183.

8.5 Ensure commercial hives/nests are disease free

- One randomised controlled trial in Canada found that the antibiotic fumagillin is not effective against *Nosema bombi* infection in managed colonies of the western bumblebee *Bombus occidentalis*. One replicated controlled trial in South Korea found that Indian meal moth *Plodia interpunctella* in commercial bumblebee colonies can be controlled with the insect pathogen *Bacillus thuringiensis* (Bt) Aizawai strain, at a strength of 1 g Bt/litre of water.

Background

Here we have summarised evidence on controlling parasites and pathogens in commercially managed bumblebees, but not in managed solitary bees or honey bees. This is because parasites and pathogens introduced in commercial bumblebee colonies are considered a serious threat to some North American bumblebee species (Otterstatter & Thomson 2008, Stout & Morales 2009).

With the possible exception of the small hive beetle (see 'Prevent introduction and spread of the small hive beetle'), the degree of threat to non-*Apis* bees from parasites of managed honey bee colonies is not yet well understood. We acknowledge that parasites and pathogens of managed honey bees, including the recently emerged invasive mite *Varroa destructor*, could have impacts on wild populations of *Apis mellifera* and other species of *Apis* in places where these species are native (such as Africa, see Dietemann *et al.* 2009). Although we have not included methods from the substantial literature on honey bee husbandry in this issue of *Bee Conservation*, we will consider including interventions to control certain honey bee parasites in future editions if they seem pertinent.

There is a growing body of literature on controlling parasites and pathogens in managed populations of solitary bees. For example, the fungus that causes chalkbrood, *Ascosphaera aggregata*, is rare in wild populations of *Megachile rotundata*, but can kill 20–50% of individuals in managed populations (Bosch & Kemp 2002, Huntzinger *et al.* 2008). Methods of controlling it have been tested, although we have not come across a recent review of this literature. Chalkbrood has been reported from wild bees (Goerzen *et al.* 1992), but we do not know of a case in which it has been suggested as a threat to species of conservation concern. If this or other pathogens emerge as a threat to declining species, we will consider including control methods in future editions.

We have captured two published studies that test methods of controlling parasites in managed bumblebee colonies.

A randomised controlled trial in a large greenhouse in Ladner, British Columbia, Canada found that the antibiotic fumagillin dicyclohexylammonium did not affect the incidence or intensity of infection by the internal parasite *Nosema bombi*, in managed colonies of the western bumblebee *Bombus occidentalis* (Whittington & Winston 2003). The antibiotic was given to 32 colonies in sugar water at doses of 26 mg/L and 52 mg/L, and 17 control colonies were not treated. The study found that samples of frass or five or more worker bees could reliably test for the presence or absence of the parasite, but could not be used to quantify the intensity of infection.

Kwon *et al.* (2003) tested methods of controlling the Indian meal moth *Plodia interpunctella*, which can be problematic to bumblebee colonies in commercial rearing conditions or greenhouses. This moth's eggs are transported in pollen from honey bee colonies, and survive normal frozen storage conditions. Replicated controlled experiments showed that storage at -60°C for 70 days killed all Indian meal moth eggs (three replicates of each treatment). Storage at -20°C killed 80–90% of the eggs. In a separate experiment, treating *B. terrestris* colonies with the insect pathogen *Bacillus thuringiensis* (Bt) Aizawai strain at a concentration of 1 g Bt/litre of water killed 98–100 % of moth larvae after seven days, but did not cause bumblebee mortality after 10 days. Treatment with 2 g Bt/litre of water caused high bumblebee mortality. There were five Bt treated colonies and five control colonies.

Bosch J. & Kemp W.P. (2002) Developing and establishing bee species as crop pollinators: the example of *Osmia* spp. (Hymenoptera: Megachilidae) and fruit trees. *Bulletin of Entomological Research*, 92, 3–16. www↗

Dietemann V., Pirk C.W.W. & Crewe R. (2009) Is there a need for conservation of honeybees in Africa? *Apidologie*, 40, 285–295.

Goerzen D.W., Dumouchel L. & Bissett J. (1992) Occurrence of chalkbrood caused by *Ascosphaera aggregata* Skou in a native leafcutting bee, *Megachile pugnata* Say (Hymenoptera, Megachilidae), in Saskatchewan. *The Canadian Entomologist*, 124, 557–558.

Huntzinger C.I., James R.R., Bosch J. & Kemp W.P. (2008) Fungicide tests on adult alfalfa leafcutting bees (Hymenoptera: Megachilidae). *Journal of Economic Entomology*, 101, 1088–1094.

Kwon Y.J., Saeed S. & Duchateau M.J. (2003) Control of *Plodia interpunctella* (Lepidoptera: Pyralidae), a pest in *Bombus terrestris* (Hymenoptera: Apidae). *The Canadian Entomologist*, 135, 893–902. www↗

Otterstatter M.C. & Thomson J.D. (2008) Does pathogen spillover from commercially reared bumble bees threaten wild pollinators? *Plos One*, 3, article e2771.

Stout J.C. & Morales C.L. (2009) Ecological impacts of invasive alien species on bees. *Apidologie*, 40, 388–409.

Whittington R. & Winston M.L. (2003) Effects of *Nosema bombi* and its treatment fumagillin on bumble bee *Bombus occidentalis* colonies. *Journal of Invertebrate Pathology*, 84, 54–58. www↗

8.6 Keep pure breeding populations of native honey bee subspecies

- One replicated trial in Switzerland found that pure breeding populations of the European black honey bee *Apis mellifera mellifera* contained a significant proportion (28%) of hybrids with an introduced subspecies *Apis mellifera carnica*.

One replicated trial estimated the degree of hybridisation in six 'pure breeding' populations of the native black honey bee *Apis mellifera mellifera*, kept by beekeepers in eastern Switzerland (Soland-Reckeweg *et al.* 2009). The introduced southeastern European subspecies *A. m. carnica* also thrives in this area. The study, based on nine honey bee genetic markers (microsatellites) and a sample of 100 black honey bee workers (a single worker from each of 100 colonies), found that 28% of the sampled bees were hybrids. In the same area, 17% of workers sampled from pure breeding populations of the introduced subspecies *A. m. carnica* were also hybrids. These findings suggest that conservation management strategies for the black honey bee need improvement, perhaps by bee breeders using genetic testing rather than conventional appearance to identify hybrids.

Soland-Reckeweg G., Heckel G., Neumann P., Fluri P. & Excoffier L. (2009) Gene flow in admixed populations and implications for the conservation of the Western honeybee, *Apis mellifera*. *Journal of Insect Conservation*, 13, 317–328. www↗

8.7 Exclude introduced European earwigs from nest sites

For other evidence relating to the use of nest boxes, including a study in areas with introduced bees, see Chapter 10 *Providing artificial nest sites for bees*.

- In California, USA, a replicated controlled trial showed that numbers of introduced European earwigs *Forficula auricularia* resting in solitary bee nest boxes can be reduced using a sticky barrier Tanglefoot. This treatment increased the use of the boxes by native bees.

Thirty drilled pine wood solitary bee nest boxes were suspended from 15 valley oak trees *Quercus lobata* on the Cosumnes River Preserve, near Sacramento, Caifornia, USA, in 1990 (Barthell *et al.* 1998). The boxes each had twelve 10 cm-deep holes, 0.65 cm in diameter. Boxes were placed in pairs. One on each tree excluded crawling earwigs *Forficula auricularia* using the sticky barrier Tanglefoot. The treatment substantially reduced the number of earwigs found in the boxes and allowed a greater total number of bee cells (during the peak bee nesting week, there were 134 cells in boxes with Tanglefoot, 45 cells in untreated boxes). The majority of nesting bees in this study were native species of the leafcutter bee genera *Megachile* and *Osmia* although introduced species of *Megachile* were also present.

Barthell J.F., Gordon W.F. & Thorp R.W. (1998) Invader effects in a community of cavity nesting Megachilid bees (Hymenoptera: Megachilidae). *Environmental Entomology*, 27, 240–247. www↗

9. Threat: problematic native species

Key Messages

Exclude bumblebee nest predators

We have captured no evidence demonstrating the effects of excluding mammalian predators from natural bumblebee nesting areas.

Exclude ants from solitary bee nesting sites

One replicated controlled trial showed that excluding ants from solitary nests of the endemic Australian bee *Exoneura nigrescens* increased the production of offspring.

9.1 Exclude bumblebee nest predators such as badgers and mink

- We have captured no evidence demonstrating the effects of excluding mammalian predators from natural bumblebee nesting areas.

9.2 Exclude ants from solitary bee nesting sites

- One replicated controlled trial showed that excluding ants from solitary nests of the endemic Australian bee *Exoneura nigrescens* increased production of offspring.

In a replicated controlled trial in Cobboboonee State Forest, Victoria, Australia, 50 single female nests of the endemic allodapine bee *Exoneura nigrescens* were protected from ants using two plastic cups and the sticky barrier Tanglefoot (Zammit *et al.* 2008). Fifty control nests were not protected. The nests, made in old flowering stems of the grass tree *Xanthorrhoea*, were set out in groups of four, one protected and one unprotected. Protected nests were more productive, with an average of 3.6 young per adult female, compared to 1.6 young per adult female in control nests.

Zammit J., Hogendoorn K. & Schwarz M. P. (2008) Strong constraints to independent nesting in a facultatively social bee: quantifying the effects of enemies-at-the-nest. *Insectes Sociaux*, 55, 74–78. www↗

10. Providing artificial nest sites for bees

Key Messages

Provide artificial nest sites for solitary bees

We have captured 30 replicated trials of nest boxes for solitary bees in 10 countries, in Europe, North and South America and Asia. Twenty-nine of them showed occupancy by bees.

Three trials on agricultural land in Germany, the USA or India showed that the number of occupied nests can double over three years with repeated nest box provision.

One small replicated trial found the number of foraging solitary bees increased in North American blueberry fields with nest boxes, compared to fields without nest boxes.

Provide artificial nest sites for bumblebees

We have found 11 replicated trials of bumblebee nest boxes. Three UK trials since 1989 showed very low uptake rates (0–2.5%) of various designs (not including underground boxes), while seven trials in previous decades in the UK, USA or Canada, and one recent trial in the USA, showed uptake rates between 10% and 48%. Seven replicated trials in the

USA, Canada or the UK have found between 6% and 58% occupancy of underground nest boxes.

We have captured no evidence for the effects of providing nest boxes on bumblebee populations.

Provide nest boxes for stingless bees

One replicated trial in Brazil found no uptake of nest boxes for stingless bees.

10.1 Provide artificial nest sites for solitary bees

See also 11.5 *Rear and manage populations of solitary bees*.

- We have captured 30 replicated trials of nest boxes for solitary bees in 10 countries, including Europe, North and South America and Asia. Twenty-nine of these trials showed occupancy by bees. Many species of solitary bee readily nest in the boxes, including some species considered endangered in a study on farmland in Germany, oil-collecting species of the genus *Centris* in South America and a recently discovered species in lowland tropical forest in Costa Rica. One trial in temperate forest in Canada recorded no bees using nest boxes.

- A set of replicated experiments in Germany estimated that four medium to large European species of solitary bee have a foraging range of 150 to 600 m, so nest boxes must be within this distance of foraging resources.

- Twenty-three replicated trials have shown nest boxes of cut hollow stems or tubes being occupied by solitary bees. Eleven trials demonstrated occupation of blocks of wood drilled with holes. Two trials in Neotropical secondary forest (one in Brazil, one in Mexico) showed that particular solitary bee species will nest in wooden boxes, without stems or confining walls inside.

- Two replicated trials have compared reproductive success in different nest box designs. One showed that reed stem and wooden grooved-board nest boxes produced more bees/nest than four other types. Nest boxes with plastic-lined holes, or plastic or paper tubes were much less productive, due to parasitism or mould. The other, a small trial, found nests of the oil-collecting bee *Centris analis* in Brazil were more productive in cardboard straws placed in drilled wooden holes than in grooved wooden boards stacked together.

- Three trials on agricultural land, one on a carpenter bee in India, one on a range of species in Germany and one on species of *Osmia* in the USA, have shown that the number of occupied solitary bee nests can double over three years with repeated nest box provision at a given site.

- One small replicated trial compared populations of solitary bees in blueberry fields in the USA with and without nest boxes over three years. The estimated number of foraging *Osmia* bees had increased in fields with nest boxes, compared to fields without nest boxes.

- Eleven replicated trials have recorded solitary bees in nest boxes being attacked by parasites or predators. Rates of mortality and parasitism have been measured in 10 studies. Mortality rates range from 13% mortality for cavity-nesting bees and wasps combined in Germany (2% were successfully parasitized), or 2% of bee brood cells attacked in shade coffee and cacao plantations in central Sulawesi, Indonesia, to 36% parasitism and 20% other mortality (56% mortality overall) for the subtropical carpenter bee *Xylocopa fenestrata* in India.

- Two replicated trials of the use of drilled wooden nest boxes by bees in California, USA, showed that introduced European earwigs *Forficula auricularia* and introduced European leafcutter bee species use the boxes. In one trial, these introduced species more commonly occupied the boxes than native bees.

- A small trial tested three soil-filled nest boxes for the mining bee *Andrena flavipes* in the UK, but they were not occupied.

Background

Solitary bee species nest either in cavities such as hollow stems or bored holes in wood or masonry, or in the ground. The provision of artificial nest sites for cavity-nesters has been widely used as a research tool, so there is a lot of literature on uptake of these nest boxes. We particularly highlight the much smaller number of studies that have looked at the effects of nest box provision on bee populations, by observing changes in bee numbers over time, preferably in areas with and without nest boxes.

We would recommend a systematic review of this subject before embarking on a strategy of providing solitary bee nest sites for conservation purposes.

Do solitary bees nest in nest boxes?

Red mason bees *Osmia rufa* readily occupied artificial nest boxes comprising metal food cans filled with drinking straws (straw diameter 5–7 mm; Free & Williams 1970). In the first year of a trial, 349 cans were recovered from 20 sites in southern England; of these 44 (13%) had one or more straws occupied by a red mason bee nest. Over the following two years, there was a tendency by this species to reoccupy cans. *Osmia caerulescens* and species of *Megachile* also occupied the cans.

The subtropical carpenter bee *Xylocopa fenestrata*, a valuable pollinator of cucurbits and other plants, has been shown to nest readily in cut stems of castor *Ricinis communis* or sarkanda *Arundo* sp. bundled together (Sihag 1993a). In a trial on agricultural land in Haryanar, India, these bees strongly preferred stems cut to 23-30 cm long, with an internal diameter of 10–12 mm. The number of occupied stems increased from 120 in the first year (1984) to 350 two years later (1986), from a total of 20,000 stems placed out over three years.

In April 1990, in Kraichgau, southwest Germany, 240 bundles of reed stems *Phragmites australis* in tins were put out, six in each of 40 fields of 10 management types, including various types of set-aside, crop fields and old meadows (Gathmann *et al.* 1994, also referred to by Tscharntke *et al.* 1998). Of 43,200 available stems, 292 were occupied by a total of 14 bee species and nine wasp species. Five species of bee considered to be endangered in Germany occupied the reed stem nests: *Anthidium lituratum*, *Heriades crenulatus*, *Megachile alpicola*, *Osmia gallarum* and *Osmia leaiana*. The two endangered *Os-*

mia species were exclusively found in nests in old meadows (more than 30 years old with several old fruit trees). The other three also nested in stems provided in 2-year-old mown set-aside, and two species (*A. lituratum* and *M. alpicola*) used reed stems in a variety of field types, including cereal crops.

Scott (1994) placed a total of 9,216 wooden nest boxes with small drilled holes of diameters 4.5–11 mm, on the edges of open fields in Upper Michigan USA, in April 1984 and 1985. Three species of the small solitary bee genus *Hylaeus* used the boxes, with an overall occupancy rate of 4%. These bees only used holes with the smallest diameters (4.5, 5.2 and 6.0 mm). *H. ellipticus* preferred the smallest 4.5 mm holes. *Hylaeus basalis* preferred nest boxes at lower heights 0.1 and 0.4 m above the ground and *H. verticalis* preferred higher boxes set at 1.1 m.

A six-year trial at two experimental farms near Poznan, western Poland demonstrated that the red mason bee *Osmia rufa* readily nests in bundles of reed stems 7–8 mm in diameter (Wójtowski *et al.* 1995).

From January 1990 to May 1991, the orchid bee *Euglossa atroveneta* made 60 nests in 50 wooden boxes placed on a table (1 m above the ground) in secondary forest planted with coffee crops, at Unión Juárez Chiapas, Mexico (Ramírez-Arriaga *et al.* 1996). The nests were made on the internal floor and walls of the boxes, constructed with resin.

In a replicated trial in central Germany from 1994–1996, 150 reed stem nest boxes (plastic tubes filled with 150×20 cm lengths of reed stem) placed at 15 different sites were occupied by 13 species of bee, 19 species of wasp and 17 species of parasite and parasitoid (Gathmann & Tscharntke 1997, also referred to by Tscharntke *et al.* 1998). In total, 8,303 nests were made.

In a replicated trial in Washington County, Maine, USA, Stubbs *et al.* (1997) added 50 drilled wooden nest boxes to each of three blueberry fields *Vaccinium angustifolium* over three years. The nest boxes each had 14 holes and were attached to trees along the field edge, at a height of 1.4 m, with 22–33 m between each box. In the first year, 30 nest boxes were occupied by bees of the genus *Osmia*, with 120 nests made. The number of nests increased the following year in all three fields. Between 3 and 11.5% of nesting holes were occupied at all three sites, each year.

Thirty to 45 drilled pine wood solitary bee nest boxes were suspended from valley oak *Quercus lobata* trees on the Cosumnes River Preserve, California, USA in 1989 and 1990 (Barthell *et al.* 1998). The boxes each had twelve 10 cm deep holes, 0.5, 0.65 or 0.8 cm in diameter. In both years, the European earwig *Forficula auricularia* was the most common occupant (59–85% of all occupied nests), followed by two introduced leafcutter bee species *Megachile rotundata* and *M. apicalis* (19.6% of all occupied nests in 1989, 3.4% in 1990). Four native bee species also occupied the boxes, but in much lower numbers. *Megachile angelarum* was found in 3.2–3.8% of occupied nests. *M. fidelis*, *M. gentilis* and *Osmia texana* occupied less than 1% of occupied nest boxes in both years.

Frankie *et al.* (1998) recorded 23 species of bee, mostly from the genera *Megachile* and *Osmia*, using nine drilled wooden nest boxes on each of six woodland, shrubland and riparian reserves over three years in northern central California, USA. Three non-native species of *Megachile* nested in the boxes – *M. apicalis*, *M. rotundata* and *M. concinna*. The former two species were common, but *M. concinna* was uncommon, recorded less than 12 times overall.

A series of four trials between 1990 and 1996 in Germany documented uptake of reed bundles placed in tins or plastic tubes attached to wooden posts (Tscharntke *et al.* 1998). Across a variety of agricultural and semi-natural habitats including orchard meadows, old hay meadows, set-aside fields, field margins and chalk grasslands, a total of 33 bee species (not including parasitic bees) used the nests.

Morato & Campos (2000) recorded 14 species of solitary bee using drilled wooden nest boxes in continuous tropical forest and inside and between forest fragments in Amazonas, Brazil. At least 108 nest boxes, each with two holes, were placed at each site. The nest boxes were more frequently occupied in continuous forest (23–29 nests/site) and natural gaps in continuous forest (78 nests/site) than in between forest fragments in pastureland or secondary vegetation (6–23 nests/site).

A trial of 120 reed stem nest boxes at 15 different agricultural sites near Göttingen in Lower Saxony, Germany in 1997 found the boxes occupied by 11 species of bee (Steffan-Dewenter 2002). The red mason bee *Osmia rufa* and the common yellow face bee *Hylaeus communis* were the most widespread and common nest box occupants in this study.

In the same study, separately reported (Gathmann & Tscharntke 2002), nest boxes had a 50% chance of being occupied by two specialised (oligolectic) species of bee – *Chelostoma rapunculi* and *Megachile lapponica* – at a distance of 256–260 m from a patch of their required forage plants. There was no colonization of nest boxes by *C. rapunculi* more than 300 m from a patch of its food plant, bellflowers *Campanula* spp..

Gathmann & Tshcarntke (2002) used translocation experiments to estimate that female solitary bees of four medium to large European species – *Andrena barbilabris, A. flavipes, A. vaga* and the red mason bee *Osmia rufa* – have a maximum foraging range between 150 to 600 m, so nest boxes have to be placed within this distance of forage resources.

From 1998–1999, Steffan-Dewenter & Leschke (2003) recorded 17,278 cells from 13 species of solitary bee using 540 reed stem nest boxes placed in 45 orchard meadows in central Germany.

In 1998, Steffan-Dewenter & Schiele (2004) placed bundles of common reed stems (153 stems per bundle, cut 15–20 cm long) in 10–13 cm diameter plastic tubes, attached to wooden posts, in orchard meadows in Germany. These were used as nest sites by the red mason bee *Osmia rufa*. Three years later, in autumn 2001, a total of 974 newly developed females were counted in 60 such nests, over five orchard meadow sites.

A study using bamboo stem nest boxes from 1994–1997 at the University of São Paulo-Ribeirão Preto, Sao Paulo, Brazil (Augusto & Garófalo 2004) recorded 5% uptake of stems by the euglossine bee *Euglossa townsendi*. A total of 383 bamboo stems were placed on outdoor shelves on a University campus, in bundles of eight to 11. Those used by female bees were 11.9 to 28.1 cm long, and 1.1–2.2 cm in internal diameter.

Three types of nest box were placed in 20 urban gardens in Sheffield, UK, from 2000–2002. They were occupied by two bee species – *Hylaeus communis* (10 gardens) and *Osmia rufa* (two gardens). The most frequently used were those constructed of 20 cm lengths of bamboo stem in plastic pipe, and 4 mm or 6 mm diameter holes drilled into wooden blocks, with uptake in over 50% of gardens over three years (Gaston *et al.* 2005). Tin cans filled with paper drinking straws (4–6 mm diameter) and 8–10 mm holes drilled in wood were less well-used.

Six different nesting materials for the red mason bee *Osmia rufa* were tested at an agricultural experimental station in Poznan County, Poland, in 2000 and 2001 (Wilkaniec & Giejdasz 2003). For each trial, 150 nests of each of the following materials were tested: reed stems, plastic tubes, paper tubes (bundles), wood, cork (grooved boards joined together in blocks), and holes drilled into wood, lined with printer acetate. All materials were used by female bees, but the highest production of bees per nest was from reed stems (3.5 bees/nest in 1999) or wood (7.2 bees/nest in 2000). Nests in paper tubes were all parasitized. Nests in plastic were well occupied (80–100%) but had a low success rate (0.2–1.8 bees/nest), partly due to mould.

Three species of wild megachilid bee (*Megachile* spp.) nested in boxes made from blocks of pine wood drilled with 14 mm long, 8 mm diameter holes, in a small replicated trial in Arizona USA (Armbrust 2004). Three sites were in the Tucson Mountains, one site on undeveloped land within Tucson city. Four nest boxes, each with 33 holes, were placed at each site. Overall 34% of available nest holes were filled between May and July 2001, but only six nests (4.5% of the available holes) were constructed in boxes at the urban site. Of the filled nests examined, 27% were subsequently occupied by parasites or predators.

Nest boxes made of 20 cm lengths of common reed *Phragmites australis* and Japanese knotweed *Reynoutria japonica*, with internal diameters from 2–20 mm, were readily occupied by the megachilid bee *Heriades fulvescens*, in a replicated trial on 24 coffee agroforestry systems in Sulawesi, Indonesia (Klein *et al.* 2004). In total, 671 nests were constructed in 240 nest boxes, over a 14-month period from 2001–2002. Four other species of Megachilidae also used these nest boxes.

A trial in secondary woodland on Santa Catarina Island, Brazil (Zillikens & Steiner 2004) showed that leafcutter bees of the species *Megachile pseudanthidioides* will nest in wooden boxes with an internal cavity (10 × 10 × 5 cm with a 10 mm diameter entrance hole), or drilled hardwood blocks (holes 7 cm long, 11–12 mm in diameter) or sections of bamboo stem (15–25 cm long, 5–25 mm in diameter).

Thiele (2002) recorded the recently discovered solitary bee *Duckeanthidium thielei* nesting in 11 and 13 mm diameter drilled holes in 24 hardwood nesting blocks placed in lowland tropical forest in Costa Rica. The species is known only from Costa Rica and considered to be rare.

In a separate report of the same study (Thiele 2005), 16 species of solitary bee were recorded nesting in 24 hardwood nesting blocks, each with 80 drilled holes, in Costa Rican lowland tropical forest. Most nests were made in boxes placed in the canopy of dead trees, 21–37 m high (69% of all nests in the first year). The author stresses the importance of retaining dead standing emergent trees for bee conservation in this habitat.

Tylianakis *et al.* (2005) placed 432 reed stem nest boxes across 48 plots in agricultural areas of Manabi Province, southwest Ecuador, on posts or hanging from trees 1.5 m above the ground. Traps were monitored from June 2003 to May 2004. In total, 31 species of bee and wasp used the traps, with averages between 8 and 12 species per plot over the entire year for different land use types. The number of bee species is not specified.

Taki *et al.* (2008) put six cardboard tube/milk carton nest boxes in each of eight forest sites in Norfolk County, southern Ontario, Canada from summer 2003 to November 2004. They recorded 12 wasp species and two species of parasitic wasp, but no bee species nesting in the boxes.

In a replicated trial in two fragments of semi-deciduous tropical forest in the State of São Paulo, Brazil, Gazola & Garófalo (2009) reported 16 species of solitary bee using nest boxes comprising bamboo stem sections or cardboard tubes. Overall, 2,708 cardboard tubes inserted in drilled wooden blocks, and 1,619 sections of bamboo cane were placed out for two years from 2000 to 2002. A total of 528 bee nests were recovered.

Oliveira & Schlindwein (2009) reported the use of nest boxes made with cardboard straws inserted into drilled wooden blocks or grooved wooden boards by the oil-collecting bee *Centris analis* in orchards in Pernambuco, Brazil. Seventeen nests were made in five wooden blocks with 40 cardboard-lined drilled holes in each (8.5% occupancy). Forty-eight nests were made in 12 nest boxes made of grooved wooden boards (10 groove holes per box) stacked together (40% occupancy). Cardboard straw nests had more brood cells (average 3.8 cells/nest) than nests in grooved boards (2.3 cells/nest). The species showed a preference for cavities with internal diameters between 6 and 10 mm.

Roubik & Villanueva-Gutiérrez (2009) monitored solitary bees using drilled wooden nesting blocks in the Sian Ka'an Biosphere Reserve, Quintana Roo, Mexico, during two 4-year stretches within a 17-year time period (1988 to 2005). Twenty $5 \times 10 \times 15$ cm blocks, each with 12 holes 7 cm long, were placed at each of five sites. The boxes were used by solitary bees from at least five genera, with the most common occupants being *Megachile zaptlana* and the oil-collecting bee *Centris analis*.

Sobek *et al.* (2009) documented three species of host bee (*Hylaeus communis*, *Hylaeus confusus* and *Megachile ligniseca*) and one parasitic bee (*Coelioxys alata*) using twelve reed stem nest boxes placed at each of 12 broadleaf woodland sites in the Hainich National Park, Thuringia, Germany. Bees made up only 9% of host cells (347 cells), with the majority of nest occupants being wasps. The abundance of nesting insects was higher in nest boxes in the canopy than in nest boxes mounted on wooden posts at chest height.

Do solitary bee nest boxes enhance local populations?

Numbers of the subtropical carpenter bee *Xylocopa fenestrata* nesting in cut stems of castor *Ricinis communis* or sarkanda *Arundo* sp. bundled together increased from 120 in the first year (1984) to 350 two years later (1986), in a trial on agricultural land in Haryanar, India (Sihag 1993a).

In a replicated trial on field margins, set aside fields and extensively managed meadows in central Germany from 1994-1996, 150 reed stem nest boxes (plastic tubes filled with 150 × 20 cm lengths of reed stem) were placed at 15 sites (Gathmann & Tscharntke 1997). The number of occupied stems almost doubled over three years from 1,761 in 1994 to 3,326 in 1996.

In a small replicated trial in Washington County, Maine, USA, Stubbs *et al.* (1997) added 50 drilled wooden nest boxes to three experimental blueberry fields *Vaccinium angustifolium* over three years from 1993 to 1995. The percentage of holes occupied rose from around 3% in the first year to over 7% in the third year in two fields, but did not rise substantially in the third field, remaining between 5 and 7%. Numbers of bees of the genus *Osmia* foraging on blueberry flowers in the experimental fields and in three control fields without nest boxes were monitored, using quadrat counts and sweep net sampling. In the first year, estimated numbers of *Osmia* ranged from 0 to 879 bees/ha in both control and experimental fields. In the third year, control fields had between 0 and 440 bees/ha, while experimental fields with nest boxes had from 219 to 1328 bees/ha. The numbers of foraging bees had increased in two of the three fields with nest boxes.

Are solitary bees in nest boxes attacked by predators and brood parasites?

From 1988-1991, Bosch (1992) recorded rates of parasitism for the orchard bee *Osmia cornuta*, nesting in nest boxes comprising paper straws in milk cartons, drilled holes in wooden blocks, grooved wooden boards, or bundles of reed stem. Overall parasitism rates ranged from 5–18% of cells in wild populations, and 0.1–13% of cells in managed populations.

Sihag (1993b) measured rates of mortality and parasitism for *X. fenestrata* nesting in stems of castor or sarkanda provided on agricultural land in Haryana, India, from 1985–1987. Up to seven generations a year were reared in stems 23–27 cm long, 10–12 mm in diameter. Cell numbers peaked in early summer (April to mid-May) and early autumn (late September to October). Mortality rates were highest in the late summer (mid-May to late-July), when parasitism was 33–36% and mortality 6–20%. Only two brood parasite species were seen, and the wasp *Monodontomerus obscurus* was responsible for 90% of the parasitism. Other larval mortality was caused mainly by ants and other predators. Birds, lizards and rodents destroyed some nests in domiciles not protected by wire cages (not quantified). No mortality caused by fungal or bacterial agents was observed.

In April 1990, in Kraichgau, southwest Germany, 240 bundles of reed stems in tins were put out, six in each of 40 fields of 10 management types, including various types of set-aside, crop fields and old meadows (Gathmann *et al.* 1994). The proportion of larvae in the nests that died from disease or failed parasitism was 13%; 2% were successfully parasitized.

In January 1990, 22% of nests of the orchid bee *Euglossa atroveneta* in wooden boxes were parasitized by the Megachilid brood parasite *Coelioxys costaricensis*, in secondary forest planted with coffee crops, at Unión Juárez Chiapas, Mexico (Ramírez-Arriaga *et al.* 1996).

Two studies between 1990 and 1996 in Germany documented predation and parasitism rates in reed bundles in tins or plastic tubes attached to wooden posts in various semi-natural and agricultural habitats (reported in Tscharntke *et al.* 1998). The percentage of bees and wasps killed by predators or parasites was 21 or 28% on average.

In a trial of 120 reed stem nest boxes in lower Saxony, Germany in 1997 (Steffan-Dewenter 2002), 14% of bee brood cells were attacked by natural enemies (brood parasites, parasitoids or predators).

Ambrust (2004) placed four nest boxes made from blocks of pine wood drilled with 14 mm long, 8 mm diameter holes at each of four sites in or near Tucson, Arizona, USA for one summer. Overall, 34% of nest holes were filled, by three species of Megachilidae. Of the filled nests examined, 27% were subsequently occupied by parasites or predators.

Klein *et al.* (2006) found 25 species of natural enemy attacking 14 bee and wasp species nesting in nest boxes in shade coffee and cacao plantations in central Sulawesi, Indonesia. 2.1% of bee brood cells were attacked in this study, although no predators or parasitoids were recorded for the most commonly found bee, the megachilid *Heriades fulvescens*, which made up 20% of all brood cells.

Gazola & Garófalo (2009) reported five and 13 species of parasite attacking bees nesting in bamboo stem and cardboard tube nest boxes respectively, in two different fragments of semi-deciduous tropical forest in the State of São Paulo, Brazil.

Sobek *et al.* (2009) recorded parasitism rates of 16% of solitary bee and wasp cells made in canopy nest boxes and 13% of cells in understorey nest boxes, during a five-month study in Hainich National Park, a semi-natural broadleaf forest in Thuringia, Germany.

Do artificial nest sites for ground-nesting bees work?

The alkali bee *Nomia melanderi*, endemic to arid and semi-arid regions of the western USA, nests in dense aggregations in patches of salty soil known as 'alkali flats'. Artificial nesting sites, or 'bee beds' have been successfully maintained for over 35 years in the United States, for the purpose of alfalfa pollination (Torchio 1987). Brief instructions for creating a bee bed are contained in Torchio's review – the bed is lined with a 40 cm deep gravel layer to hold water, covered with 1 m of soil mixed with salt and seeded with soil cores from active nesting areas. We have not sought evidence on the take-up rates, or effects of management regimes on these artificial nest sites, because the alkali bee is a managed pollinator with very specific requirements.

A small-scale study with three replicates tested soil filled nest boxes for the mining bee *Andrena flavipes*, a host of the Nationally Scarce dotted bee-fly *Bombylius discolor* in the UK. These nest boxes were not occupied, despite being placed alongside active colonies of nesting bees (Gibbs 2004).

Armbrust E.A. (2004) Resource use and nesting behaviour of *Megachile prosopidis* and *M. chilopsidis* with notes on *M. dischorina* (Hymenoptera: Megachilidae). *Journal of the Kansas Entomological Society*, 77, 89–98. www↗

Augusto S.C. & Garófalo C.A. (2004) Nesting biology and social structure of *Euglossa (Euglossa) townsendi* Cockerell (Hymenoptera, Apidae, Euglossini). *Insectes Sociaux*, 51, 400–409. www↗

Barthell J.F., Gordon W.F. & Thorp R.W. (1998) Invader effects in a community of cavity nesting Megachilid bees (Hymenoptera: Megachilidae). *Environmental Entomology*, 27, 240–247. www↗

Bosch J. (1992) Parasitism in wild and managed populations of the almond pollinator *Osmia cornuta* Latr. (Hymenoptera: Megachilidae). *Journal of Apicultural Research*, 31, 77–82. www↗

Frankie G.W., Thorp R.W., Newstrom-Lloyd L.E., Rizzardi M.A., Barthell J.F., Griswold T.L., Kim J-Y. & Kappagoda S. (1998) Monitoring solitary bees in modified wildland habitats: implications for bee ecology and conservation. *Environmental Entomology*, 27, 1137–1148. www↗

Free J.B. & Williams I.H. (1970) Preliminary investigations on the occupation of artificial nests by *Osmia rufa* L. (Hymenoptera, Megachilidae). *Journal of Applied Ecology*, 73, 559–566. www↗

Gaston K.J., Smith R.M., Thompson K. & Warren P.H. (2005) Urban domestic gardens (II): experimental tests of methods for increasing biodiversity. *Biodiversity and Conservation*, 14, 395–413. www↗

Gathmann A. & Tscharntke T. (1997) Bienen und Wespen in der Agrarlandschaft (Hymenoptera Aculeata): Ansiedlung und Vermehrung in Nisthilfen [Bees and wasps in the agricultural landscape (Hymenoptera Aculeata): colonization and augmentation in trap nests]. *Mitteilungen der Deutschen Gesellschaft für Allgemeine und Angewandte Entomologie*, 11, 91–94. www↗

Gathmann A. & Tscharntke T. (2002) Foraging ranges of solitary bees. *Journal of Animal Ecology*, 71, 757–764. www↗

Gathmann A., Greiler H-J. & Tscharntke T. (1994) Trap-nesting bees and wasps colonizing set-aside fields: succession and body size, management by cutting and sowing. *Oecologia*, 98, 8–14. www↗

Gazola A.L. & Garófalo C.A. (2009) Trap-nesting bees (Hymenoptera: Apoidea) in forest fragments of the State of São Paulo, Brazil. *Genetics and Molecular Research*, 8, 607–622. www↗

Gibbs D. (2004) The dotted bee-fly (*Bombylius discolor* Mikan 1796). A report on the survey and research work undertaken between 1999 and 2003. English Nature Research Report 583. www↗

Klein A.M., Steffan-Dewenter I. & Tscharntke T. (2004) Foraging trip duration and density of megachilid bees, eumenid wasps and pompilid wasps in tropical agroforestry systems. *Journal of Animal Ecology*, 73, 517–525. www↗

Klein A.M., Steffan-Dewenter I. & Tscharntke T. (2006) Rain forest promotes trophic interactions and diversity of trap-nesting Hymenoptera in adjacent agroforestry. *Journal of Animal Ecology*, 75, 315–323. www↗

Morato E.F. & Campos L.A.O. (2000) Efeitos da fragmentação florestal sobre vespas e abelhas solitárias em uma área da Amazônia Central [Effects of forest fragmentation on solitary wasps and bees in an area of central Amazonia]. *Revista Brasileira de Zoologia*, 17, 429–444. www↗

Oliveira R. & Schlindwein C. (2009) Searching for a manageable pollinator for acerola orchards: the solitary oil-collecting bee *Centris analis* (Hymenoptera: Apidae: Centridini). *Journal of Economic Entomology*, 102, 265–273. www↗

Ramírez-Arriaga E., Cuadriello-Aguilar J.I. & Martínez Hernández E. (1996) Nest structure and parasite of *Euglossa atroveneta* Dressler (Apidae: Bombinae: Euglossini) at Unión Juárez, Chiapas, México. *Journal of the Kansas Entomological Society*, 69, 144–152. www↗

Roubik D.W. & Villanueva-Gutiérrez R. (2009) Invasive Africanized honey bee impact on native solitary bees: a pollen resource and trap analysis. *Biological Journal of the Linnaean Society*, 98, 152–160. www↗

Scott V.L. (1994) Phenology and trap selection of three species of *Hylaeus* (Hymenoptera: Colletidae) in Upper Michigan. *The Great Lakes Entomologist*, 27, 39–47. www↗

Sihag R.C. (1993a) Behaviour and ecology of the subtropical bee *Xylocopa fenestrata* F. 7. Nest preferences and response to nest translocation. *Journal of Apicultural Research*, 32, 102–108. www↗

Sihag R.C. (1993b) Behaviour and ecology of the subtropical bee *Xylocopa fenestrata* F. 8. Life cycle, seasonal mortality, parasites and sex ratio. *Journal of Apicultural Research*, 32, 109–114. www↗

Sobek S., Tscharntke T., Scherber C., Schiele S. & Steffan-Dewenter I. (2009) Canopy vs. understorey: Does tree diversity affect bee and wasp communities and their natural enemies across forest strata? *Forest Ecology and Management*, 258, 609–615. www↗

Steffan-Dewenter I. (2002) Landscape context affects trap-nesting bees, wasps, and their natural enemies. *Ecological Entomology*, 27, 631–637. www↗

Steffan-Dewenter I. & Leschke K. (2003) Effects of habitat management on vegetation and above-ground nesting bees and wasps of orchard meadows in Central Europe. *Biodiversity and Conservation*, 12, 1953–1968. www↗

Steffan-Dewenter I. & Schiele S. (2004) Nest site fidelity, body weight and population size of the red mason bee, *Osmia rufa* (Hymenoptera: Megachilidae), evaluated by mark-recapture experiments. *Entomologia Generalis*, 27, 123–131. www↗

Stubbs C.S., Drummond F.A. & Allard S.L. (1997) Bee conservation and increasing *Osmia* spp. in Maine lowbush blueberry fields. *Northeastern Naturalist*, 4, 133–144. www↗

Taki H., Viana B.F., Kevan P.G., Silva F.O. & Buck M. (2008) Does forest loss affect the communities of trap-nesting wasps (Hymenoptera: Aculeata) in forests? Landscape vs. local habitat conditions. *Journal of Insect Conservation*, 12, 15–21. www↗

Thiele R. (2002) Nesting biology and seasonality of *Duckeanthidium thielei* Michener (Hymenoptera: Megachilidae), an oligolectic rainforest bee. *Journal of the Kansas Entomological Society*, 75, 274–282. www↗

Thiele R. (2005) Phenology and nest site preferences of wood-nesting bees in a Neotropical lowland rain forest. *Studies on Neotropical Fauna and Environment*, 40, 39–48. www↗

Torchio P. (1987) Use of non-honey bee species as pollinators of crops. *Proceedings of the Entomological Society of Ontario*, 118, 111–124.

Tscharntke T., Gathmann A. & Steffan-Dewenter I. (1998) Bioindication using trap-nesting bees and wasps and their natural enemies: community structure and interactions. *Journal of Applied Ecology*, 35, 708–719. www↗

Tylianakis J.M., Klein A.M. & Tscharntke T. (2005) Spatiotemporal variation in the diversity of hymenoptera across a tropical habitat gradient. *Ecology*, 86, 3296–3302. www↗

Wilkaniec Z. & Giejdasz K. (2003) Suitability of nesting substrates for the cavity-nesting bee *Osmia rufa*. *Journal of Apicultural Research*, 42, 29–31. www↗

Zillikens A. & Steiner J. (2004) Nest architecture, life cycle and cleptoparasite of the Neotropical leaf-cutting bee *Megachile (Chrysosarus) pseudanthidioides* Moure (Hymenoptera: Megachilidae). *Journal of the Kansas Entomological Society*, 77, 193–202. www↗

10.2 Provide artificial nest sites for bumblebees

- We have captured 11 replicated trials of bumblebee nest boxes. Several different types of nest box have been shown to be acceptable to bumblebees, including wooden or brick and tile boxes at the ground surface, underground tin, wooden or terracotta boxes and boxes attached to trees.

- Three replicated trials since 1989 in the UK have shown very low uptake rates (0–2.5%) of various nest box designs (not including underground nest boxes), while seven trials in previous decades in the UK, USA or Canada, and one recent trial in the USA, showed uptake rates between 10% and 48%.

- Wooden surface or above ground nest boxes of the kind currently marketed for 'wildlife gardening' are not the most effective design. Eight studies test this type of nest box. Five (pre-1978, USA or Canada) find 10–40% occupancy. Three (post-1989, UK) find very low occupancy of 0–1.5%. The four replicated trials that have directly compared wooden surface nest boxes with other types all report that underground, false underground or aerial boxes are more readily occupied.

- Nest boxes entirely buried 5–10 cm underground, with a 30–80 cm long entrance pipe, are generally the most effective. Seven replicated trials in the USA, Canada or the UK have tested underground nest boxes and found between 6% and 58% occupancy.

- We have captured no evidence for the effects of providing nest boxes on bumblebee populations.

Sladen (1912) placed 112 underground nest boxes for bumblebees in his garden near Dover, in Kent, England in 1910 and 1911. Boxes were buried cylinders of tin or terracotta, or holes in the ground with a wooden cover, and a 38–75 cm tunnel leading to them. They were occupied by six species of bumblebee, including the short-haired bumblebee *Bombus subterraneus* now extinct from Britain. Thriving colonies developed in 13–19% of nest sites provided.

A trial of 36 underground bumblebee nest boxes in woodland and meadows near Urbana, Illinois, USA, found 48% of the boxes were occupied by a total of five species of bumblebee from 1915 to 1919 (Frison 1926). The boxes were made of tin or cypress wood, provided with grass from field mouse nests, and had an entrance spout or pipe at ground level. Some had a copper gauze base, to allow drainage.

A trial of 172 nest boxes of six types (unequally replicated), carried out on farms in Wisconsin, USA in 1953, showed that bumblebees will nest in wooden nest boxes or half-buried flower pots at the surface, wooden boxes attached to buildings 1 m above ground, or in metal cans or roof tile enclosures buried underground (Fye & Medler 1954). Flax straw, old mouse nests or felt were added as bedding. Overall, 34% of the nest boxes were occupied, by five species of bumblebee, including three now thought to be declining in some parts of North America (Xerces Society 2008): the red-belted bumblebee *Bombus rufocinctus*, the yellow bumblebee *B. fervidus* and the half-black bumblebee *B. vagans*.

A trial of 500 above ground wooden nest boxes near Lethbridge in southern Alberta, Canada, found that over 10% of boxes placed in uncultivated gardens, beside fence posts on prairie, or along copses were used (Hobbs *et al.* 1960). Upholsterer's cotton was used as bedding. Boxes placed in long grass were not used. Seven species used the nest boxes, including *B. rufocinctus* and *B. fervidus*, both thought to be declining in parts of North America, and the Western bumblebee *B. occidentalis* (one nest only), which has undergone dramatic range contraction recently (*B. occidentalis* may be a Western variant of another species *Bombus terricola* rather than a species in its own right - see www.nhm.ac.uk/research-curation/research/projects/bombus/bo.html). Two important alfalfa crop pollinators in Alberta – the yellow-banded bumblebee *B. terricola* and the red-belted or tri-colored bumblebee *B. ternarius* – did not use the boxes.

A trial of 1,023 wooden nest boxes placed in grassland or woodland in southern Alberta, Canada (Hobbs *et al.* 1962) found an occupancy rate by bumblebees of 35% overall. Underground nest boxes were more often occupied (49%) than above ground (32%) or half-buried (36%) boxes.

A trial of 1,233 surface boxes, 465 underground boxes, 500 false underground boxes and 100 above ground boxes in areas of mixed woodland and grassland in southern Alberta, Canada, from 1961 to 1966 (Hobbs 1967), found underground and false underground boxes were more often occupied by bumblebees (approximately 58% and 48% respectively) than surface boxes (approximately 26%) or above ground boxes attached to tree trunks (35%). False underground boxes were at the surface, but with a partially buried entrance pipe giving the appearance of a subterranean nest.

A replicated trial carried out in 1970 and 1971 in southwestern Alberta, Canada, found that 23% and 43% of wooden nest boxes put out for bumblebees were occupied, in the respective years (Richards 1978). In total, 2,140 boxes were put out in a 1 km^2 area, with equal numbers of underground, false underground, surface and above ground boxes. Upholsterer's cotton was added to each box as bedding. Fourteen different species of bumblebee *Bombus* sp. used the boxes. Preferred nest box locations were underground, buried 10 cm below the surface with a 30 cm plastic pipe to the entrance (38.5% occupied), and above ground, with the box wired to a tree trunk at chest height (38.7% occupied). False underground and surface nest boxes were also readily occupied (22.6% and 32.7% respectively).

A trial (unequally replicated) of 654 bumblebee nest boxes over three years (1989-1991) in farmland, gardens and fenland in Cambridgeshire, UK, found only 10 boxes were occupied (1.5%) (Fussell & Corbet 1992). The nest boxes tested were wooden boxes raised 10 cm or 1 m above ground, or nest sites constructed with bricks and concrete tiles on the ground. Dry moss, felt or shredded textiles were added as bedding. Two common and widespread bumblebee species used boxes of both types, the early bumblebee *Bombus pratorum* and the common carder bee *B. pascuorum*.

During a three-year study in Sheffield, UK, no artificial nest chambers of any design (above ground terracotta plant pots, buried terracotta plant pots with entrance holes at the top (no pipe) and wooden boxes) were occupied by bumblebees *Bombus* spp. (Gaston *et al.* 2005). Between 52 and 72 nest boxes were put out in each year, in 20 domestic gardens.

Elliott (2009) reports putting out 100 wooden nest boxes in subalpine meadows in Gunnison National Forest, Colorado, USA, of which approximately 10% were occupied by the *Bombus appositus*, a long tongued bumblebee and one of the three most abundant bumblebee species in the study area. These nest boxes were lined with cotton for insulation, but no further detail of their design is given.

Lye (2009) tested six different bumblebee nest box designs in gardens and farmland in England and Scotland: aerial wicker nest boxes (120), dug holes covered with concrete slabs or upturned flower pots (100 each), semi-underground wooden nest boxes (100), wooden surface boxes (26) and a buried nest box design made with two pairs of flower pots placed mouth to mouth (170). She found very low uptake rates of 0–2% for all designs except the underground flowerpot design, which incorporated drainage, ventilation and a 30 cm entrance pipe. For this design, 2% of 150 were used on Scottish farmland, but 40% (eight of 20) of those put out in an English botanic garden supported bumblebee colonies. Two of 20 aerial wicker nest boxes (10%) were occupied at the same site and one of 100 placed at a site in Scotland.

We are aware of at least three studies of bumblebee nest boxes in New Zealand, where bumblebees were introduced from the UK (Donovan & Weir 1978, Pomeroy 1981, Macfarlane *et al.* 1983). These studies find occupancy rates of 8–88% for different nest box designs, with the highest occupancy rate (88%) recorded for underground nest boxes in one study (reviewed in Lye 2009). They are not summarised by Conservation Evidence, because providing nest boxes for non-native and potentially invasive species is not a conservation intervention.

Donovan B.J. & Weir S.S. (1978) Development of hives for field population increase, and studies on the life cycle of the four species of introduced bumble bees in New Zealand. *New Zealand Journal of Agricultural Research*, 21, 733–756.

Elliott S.E. (2009) Surplus nectar available for subalpine bumble bee colony growth. *Environmental Entomology*, 38, 1680–1689. www↗

Frison T.H. (1926) Experiments in attracting queen bumblebees to artificial domiciles. *Journal of Economic Entomology*, 19, 149–155. www↗

Fussell M. & Corbet S. (1992) The nesting places of some British bumble bees. *Journal of Apicultural Research*, 31, 32–41. www↗

Fye R.E. & Medler J.T. (1954) Field domiciles for bumblebees. *Journal of Economic Entomology*, 47, 672–676. www↗

Gaston K.J., Smith R.M., Thompson K. & Warren P.H. (2005) Urban domestic gardens (II): experimental tests of methods for increasing biodiversity. *Biodiversity and Conservation*, 14, 395–413. www↗

Hobbs G.A., Virostek J.F. & Nummi W.O. (1960) Establishment of *Bombus* spp. (Hymenoptera: Apidae) in artificial domiciles in Southern Alberta. *The Canadian Entomologist*, 92, 868–872. www↗

Hobbs G.A., Nummi W.O. & Virostek J.F. (1962) Managing colonies of bumble bees (Hymenoptera: Apidae) for pollination purposes. *The Canadian Entomologist*, 94, 1121–1132. www↗

Hobbs G.A. (1967) Obtaining and protecting red-clover pollinating species of *Bombus* (Hymenoptera: Apidae). *The Canadian Entomologist*, 99, 943–951. www↗

Lye G. (2009) Nesting ecology, management and population genetics of bumblebees: an integrated approach to the conservation of an endangered pollinator taxon. PhD thesis, Stirling University. www↗

MacFarlane R.P., Griffin R.P. & Read P.E.C. (1983) Bumble bee management options to improve 'grasslands pawera' red clover seed yields. *Proceedings of the New Zealand Grasslands Association*, 44, 47–53.

Pomeroy N. (1981) Use of natural sites and field hives by a long-tongued bumble bee *Bombus ruderatus. New Zealand Journal of Agricultural Research*, 24, 409–414.

Richards K.W. (1978) Nest site selection by bumble bees (Hymenoptera: Apidae) in Southern Alberta. *The Canadian Entomologist*, 110, 301–318. www↗

Sladen F.W.L. (1912) *The Humble Bee: its Life History and How to Domesticate it.* Macmillan and Co, London. www↗

Xerces Society (2008) *Bumblebees in decline. Invertebrate conservation fact sheet.* Available at http://www.xerces.org/wp-content/uploads/2008/09/bumblebees_factsheet.pdf. Accessed 2 December 2009.

10.3 Provide nest boxes for stingless bees

See Chapter 6.2 *Replace honey-hunting with apiculture* for evidence on methods of stingless bee-keeping.

• One replicated trial tested nest boxes placed in trees for the stingless bee *Melipona quadrifasciata* in Brazil and found no uptake.

Antonini & Martins (2003) erected 40 nest boxes (25 × 25 × 40 cm) for stingless bees on tree branches 3–5 m above ground, in pristine and degraded cerrado (grass/shrubland) in Minas Gerais, Brazil, in March 1999. None were occupied by any stingless bee colonies, although 48 natural nests were found in the 18 km² study area. The lack of nest box uptake was thought to be due to an abundance of natural nest sites.

Antonini Y. & Martins R.P. (2003) The value of a tree species (*Caryocar brasiliense*) for a stingless bee *Melipona quadrifasciata quadrifasciata. Journal of Insect Conservation*, 7, 167–174. www↗

11. Captive breeding and rearing of wild bees (ex-*situ* conservation)

Key Messages

Rear declining bumblebees in captivity

We have found 22 trials documenting captive rearing of bumblebee colonies from 13 countries in Europe, North and Central America and Asia. Amongst these are trials that reared bumblebee species now declining in parts of North America (*Bombus terricola*) or the UK (*B. ruderatus*).

Re-introduce laboratory-reared bumblebee queens to the wild

We have found no evidence for the effects of reintroducing queens.

Re-introduce laboratory-reared bumblebee colonies to the wild

Seven replicated trials have monitored the success of laboratory-reared colonies of bumblebees introduced to the environment in Europe or North America. In four of these (three in the UK, 1 in Canada) colonies were allowed to develop until new queens were produced. In two, the numbers of queens/colony were very low or zero and in the other two, good numbers of new queens were produced.

Translocate bumblebee colonies in nest boxes

Three small trials in Canada or the UK have tested the effect of translocating bumblebee colonies in nest boxes. Just one, a UK trial, concluded that early bumblebee *Bombus pratorum* colonies adapt well to being moved.

Rear and manage populations of solitary bees

Several species of solitary bee are reared and managed commercially as pollinators. These species readily nest in drilled holes or stacked grooved boards of wood or polystyrene. Three management trials in the USA or Poland with megachilids not commercially managed, and a review of studies of managed species, found that local populations can increase up to six-fold in one year under management, if conditions are good and plentiful floral resources are provided.

Translocate solitary bees

One replicated trial in India showed that translocating solitary bees in immature stages, but not as adults, can result in establishment of populations at new sites.

Introduce mated females to small populations to improve genetic diversity

One trial in Brazil showed that genetic diversity can be maintained in small isolated populations of social bees by regularly introducing inseminated queens.

11.1 Rear declining bumblebees in captivity

For control of bumblebee predators and parasites in artificial rearing conditions, see section: Chapter 8.5 *Ensure commercial hives/nests are disease free*.

- We have captured 22 trials from 13 countries documenting captive rearing of bumblebee colonies by confining mated queens alone (eight trials), with one or more bumblebee workers (seven trials), honey bee workers (one trial), male bumblebee pupae (three trials) or following anaesthetisation with CO_2 (four trials). One trial found that over four years of artificial rearing, *Bombus terrestris* queens gradually decreased in weight.

- Three trials have tried to rear North American bumblebees now declining or thought to be declining. Two induced spring queens of the half-black bumblebee

B. vagans to rear adults in captivity, one trial induced queen yellow-banded bumblebees *B. terricola* (attempted in all three trials) and red-belted bumblebees *B. rufocinctus* (only attempted in one trial) to rear adults in captivity. All three trials tried to rear the yellow bumblebee *B. fervidus* and in all cases the queens laid eggs but the larvae died before becoming adults. One trial found the same pattern for the rusty-patched bumblebee *B. affinis* and the American bumblebee *B. pensylvanicus*. One study reports rearing the large garden bumblebee *B. ruderatus*, a Biodiversity Action Plan species in the UK.

• Two trials have reported laboratory rearing of a pocket-making bumblebee, the Neotropical *B. atratus*.

• Three replicated trials demonstrated that the pollen diet of captively reared bumblebees influences reproductive success. In one trial, buff-tailed bumblebee *B. terrestris* colonies fed on freshly frozen pollen produced larger queens that survived better and produced larger colonies themselves than colonies fed on dried, frozen pollen. Two replicated trials demonstrated that *B. terrestris* workers can produce more offspring when fed types of pollen with a higher protein content.

• Two replicated experiments showed that an artificial light regime of eight hours light, 16 hours darkness, can reduce the time taken for artificially reared queen *B. terrestris* to lay eggs, relative to rearing in constant darkness.

• We have captured two replicated trials examining the effect of different artificial hibernation regimes in *B. terrestris*. One showed that hibernation of queens at 4–5°C for 45 days enhanced egg-laying and colony formation rates, but hibernated queens produced smaller colonies than non-hibernated queens. The second showed that queens should weigh more than 0.6 g (wet weight) and be hibernated for four months or less to have a good chance of surviving.

Background

Methods for rearing bumblebee colonies in the laboratory have developed substantially in recent years. Since 1987, bumblebees have been reared commercially all year round for pollination purposes. The most widely sold species are the buff-tailed bumblebee *Bombus terrestris* in Eurasia and the common eastern bumblebee *B. impatiens* in the USA, but the white-tailed bumblebee *B. lucorum*, *B. ignitus* and the Western bumblebee *B. occidentalis* have also been made available (Velthuis & van Doorn 2006). *B. occidentalis* has declined dramatically within its former range in western North America.

In commercial operations, mated queens are stored between 1° and 5°C until needed, usually after a one-week transition period at an intermediate temperature. Then they may receive a CO_2 anaesthetic to induce egg-laying, before being confined in nest boxes with one or more conspecific workers or honey bee *Apis mellifera* workers. Standard rearing conditions are complete darkness, 28 (± 1)°C and 60 (± 5)% relative humidity, regularly supplied with freshly frozen pollen collected by honey bees and sugar syrup in the ratio 1:1 with water, by volume.

This process works best with bumblebees of the type called 'pollen-storers', whose workers accept pollen placed anywhere near the brood and feed larvae individually. In the other type of bumblebee, 'pocket-makers' (section Odontobombus), which often have longer tongues, larvae largely feed themselves and workers deliver pollen to pockets beneath the brood clump. None of the commercially reared bumblebee species is a pocket-maker.

The ITIS world bee checklist lists 258 species of bumblebee (genus *Bombus*). Seven are considered Critically Endangered, Endangered or Vulnerable globally (Williams & Osborne 2009) and many others are of concern or suspected to be declining in parts of Europe or North America (Xerces 2008).

If captive rearing of rare and declining bumblebee species becomes a popular intervention, we recommend a systematic review of methods. We are aware of other studies in this area that are difficult to access or require translation.

How can queen bumblebees be induced to form colonies in captivity?

Sladen (1912) reared more than eight colonies of the buff-tailed bumblebee *Bombus terrestris* and one of the red-tailed bumblebee *B. lapidarius* by confining one or two nest searching queens with between two and seven workers of the same species in wooden boxes supplied with honey and pollen. In the case of the red-tailed bumblebee, the queen was also confined with clusters of cocoons from another nest.

Plath (1923) induced six different species of native North American bumblebee queens to lay eggs and rear colonies of adults, by confining spring queens with one to three bumblebee workers in dark wooden nest boxes supplied with honey bee-collected pollen and diluted honey. These six species included the half-black bumblebee *B. vagans*, thought to be declining in the USA. Five other species treated the same way laid eggs but did not rear colonies because the larvae died. Four of the species that could not be reared are also declining or thought to be declining in the United States: the rusty-patched bumblebee *B. affinis*, the yellow-banded bumblebee *B. terricola*, the American bumblebee *B. pensylvanicus* and the yellow bumblebee *B. fervidus*. The latter two species are reported to be 'pocketmakers' (Kearns & Thomson 2001).

Frison (1927) induced nine different species of native North American bumblebee queens to lay eggs, by confining 'broody' spring queens (already secreting wax) singly or in pairs, in wooden boxes in the dark. Fresh honeybee pollen and diluted honey solution were supplied. Colonies were reared to produce adult workers in 11 of the 46 trials between 1917 and 1920, including by the half-black bumblebee *B. vagans*, thought to be declining in the USA. Two other species reported to be declining: the yellow bumblebee *B. fervidus* and the yellow-banded bumblebee *B. terricola* were induced to lay eggs but did not rear colonies. The larvae died. No eggs were laid in two experiments with the American bumblebee *B. pensylvanicus*.

In Sweden, Hasselrot (1952) induced hibernated spring queens of three bumblebee species (*B. terrestris*, the tree bumblebee *B. hypnorum* and *B. lapidarius*) to form colonies in 26 out of 30 wooden nest boxes. He confined them alone in two linked boxes and provided honey solution and fresh pollen, moss and cellulose nesting material.

Plowright & Jay (1966) induced mated queens of seven Canadian bumblebee species to found colonies in captivity, by confining them singly or in pairs in wooden boxes kept at 21° or 29° C, regularly provided with fresh pollen and honey solution. Twenty-eight of the 30 B. terricola tested and four of the nine red-belted bumblebees B. rufocinctus reared adults using this method, but a single yellow bumblebee B. fervidus did not. Some queens confined in waxed paper cartons laid eggs, but none successfully reared adult workers.

Pomeroy & Plowright (1980) described two hive designs in which they had reared several species of bumblebee in the laboratory, including the pocket-making Neotropical species B. atratus and the large garden bumblebee B. ruderatus. Both designs were internally cone-shaped – one made of metal, one moulded from porous concrete. The metal hive had a heating element, and its internal temperature controlled at around 30°C by thermostat.

In a replicated trial in Germany, Röseler (1985) demonstrated that mated queen B. terrestris can be induced to lay eggs by anaesthestising them with CO_2 for 30 minutes on two consecutive days. After this treatment, 73% of unhibernated and 81% of hibernated queens began egg-laying within one week of confinement with workers.

Two replicated laboratory trials in France (Tasei & Aupinel 1994, Tasei 1994) showed that an eight hour light, 16 hour dark regime imposed during rearing induces egg-laying more quickly (average 33 and 21 days respectively) in B. terrestris queens than constant dark (average 47 and 39 days), or, in one set of experiments, constant light (average 59 days to egg-laying). Both experiments found that the light:dark regime did not significantly affect the chance of a B. terrestris queen laying eggs (range 61–73% for all treatments). Tasei & Aupinel (1994) used 103 artificially hibernated laboratory reared queens confined alone in standard rearing conditions. Tasei (1994) used 200 non-hibernated laboratory reared queens anaesthetized with CO_2 and confined with one B. terrestris worker.

Beekman et al. (1998) tested the effects of different artificial hibernation regimes (temperatures from -5 to 15°C, durations from 1 to 8 months) on 2,210 queen B. terrestris from laboratory-reared colonies in the Netherlands. A queen's initial weight and the duration of hibernation strongly affected survival, but the temperature did not. Queens should weigh more than 0.6 g (wet weight) and be hibernated for four months or less to have a good chance of surviving. Queens weighing less than 0.6 g before hibernation did not survive, but above this threshold, initial weight did not affect survival. Few queens survived hibernation periods of 6 and 8 months (8.5%, compared to 83% of queens hibernated for one, two and four months). Neither temperature, weight nor length of hibernation affected a queen's ability to lay eggs after surviving hibernation.

Beekman *et al.* (2000) reared *B. terrestris* in the laboratory over four years, with one to three generations per year, starting with the progeny of 47 wild-caught queens in 1993. A total of 170 colonies were reared altogether. Queens were mated in mating cages, hibernated for two to four months and induced to form colonies by confining with two to four honey bee workers in standard rearing conditions. They found a significant linear decrease in average queen weight over time, from 0.83 g in 1993 to 0.73 g in 1996. Since queens weighing less than 0.6 g do not survive hibernation, this change would be of concern in the context of captive-bred releases. Beekman *et al.*'s experimental results suggest it is caused by a nutrient deficiency, rather than inbreeding or reallocation of resources within colonies.

Yeninar *et al.* (2000) reared 96 colonies of the Mediterranean subspecies *B. terrestris dalmatinus* from nest-searching queens caught in the wild in Turkey. This Mediterranean subspecies aestivates over the dry season from June to October, rather than hibernating over winter. It is the subspecies most commonly reared commercially.

Queens were confined in standard rearing conditions with a single male *B. terrestris* pupa. They produced an average of 152 workers, 258 males and 31 queens each, but 48% of the colonies produced no queens.

In experiments in South Korea with 132 field-caught, naturally hibernated *B. ignitus* queens, Yoon *et al.* (2002) found 27°C and 65% relative humidity produced higher rates of colony foundation and better colony performance than other temperatures and levels of humidity. Queens were confined alone to induce colony formation. At 27°C, 83% of queens founded colonies, 63% produced colonies with more than 50 workers and 46% produced new queens. These percentages were 2.2–5.5 times higher than rates achieved at 23° and 30°C. Colonies reared at 27°C produced more workers and more queens than those at other temperatures, and developed two to five times faster.

Kwon *et al.* (2003) found that confining 100 *B. terrestris* queens with a young male pupa 1–2 days old stimulated egg laying and improved overall colony productivity compared to queens that had been confined with older pupae (9–11 days old). Eighty per cent of the 20 queens given a 1–2 day old pupa produced a colony, compared to 30% of 20 queens given a 9–11 day old pupa.

Lopez-Vaamonde *et al.* (2004) reared colonies from wild-caught queens of the native UK subspecies *B. terrestris audax*. Rearing methods are not given in detail, but 32 colonies of at least 10 workers were reared from 122 queens.

Ings *et al.* (2006) reared wild-caught queens of the native UK subspecies *B. terrestris audax* in southern England. Queens were confined in dual compartment nest boxes at 25–28°C, 60% relative humidity, with two or three male pupae or artificial pupae made of clay. From 79 nest searching queens caught in March, 20 colonies were reared to the second brood of workers.

Almanza (2007) reported rearing four colonies of the Neotropical pocket-making species *B. atratus* from wild-caught queens in Colombia. Kept alone under standard rearing conditions, the queens began egg-laying within one week. The colonies lasted between 97 and 183 days, and produced between 40 and 145 workers, but none produced any new queens or males.

Gurel & Gosterit (2008) found *B. terrestris* queens were more likely to lay eggs and found colonies, and laid eggs more quickly, when they had been confined with a single *B. terrestris* worker, compared to queens confined alone, with a honey bee worker or with a *B. terrestris* pupa. These laboratory experiments were carried out at the University of Akdeniz, Turkey, following hibernation and CO_2 anaesthetic treatment of laboratory-reared queens.

Li *et al.* (2008) reared colonies of two bumblebee species native to China – the white-tailed bumblebee *B. lucorum* and *B. ignitus* – from queens caught in the field in May (rearing methods not described in detail); 84% of 150 *B. ignitus* queens, and 89% of 200 *B. lucorum* queens laid eggs. Colonies produced 105–107 workers/colony on average, with no difference between species, but *B. lucorum* colonies produced significantly more queens (average 121 queens/colony) than *B. ignitus* (average 55 queens/colony).

Gurel & Gosterit (2009) reared 50 wild-caught *B. terrestris dalmatinus* over two generations in the laboratory. Queens were anaesthetized with CO_2, then confined with *B. terrestris* workers to induce colony formation. They found second-generation queens produced around 60% more workers (average 121 workers/colony, compared to 72 workers/colony in the first generation), significantly more males (average 71 males/colony, compared to 30 for first-generation colonies) and completed the colony cycle significantly more quickly than first-generation colonies.

In another replicated controlled laboratory experiment, Gosterit & Gurel (2009) found that hibernating *B. terrestris* queens at 4–5°C for 45 days followed by anaesthetizing with CO_2 for 30 minutes produced the highest egg-laying and colony formation rates, compared to non-hibernated queens, or those hibernated for 75 or 105 days. However, non-hibernated queens (also anaesthetized) ultimately produced larger colonies, with more workers and more new queens and males. These experiments used 148 mated, laboratory reared queens, with 30–40 queens in each treatment group.

Chiang *et al.* (2009) documented rearing of two montane oriental species, *B. eximius* and *B. sonani* in Taiwan. Queens were induced to form colonies by confining them alone in wooden boxes at 26°C and 65% relative humidity, under red light. Of 53 *B. eximius* queens, 40 (76%) laid eggs, and 31 produced mature colonies. Of 37 *B. sonani* queens, 27 (73%) laid eggs and 22 produced mature colonies. *B. eximius* produced significantly larger colonies with on average 120 workers, 210 males and 25 queens, compared to 53 workers, 102 males and nine queens on average for *B. sonani*.

Whitehorn *et al.* (2009) reared colonies of *B. terrestris* from 210 commercially-reared queens by confining queens alone, under standard rearing conditions, following artificial hibernation for 47 days at 6°C. Ninety-three queens (44%) survived artificial hibernation and 47 of them (51% of those surviving hibernation) founded colonies with at least one offspring.

Does pollen diet affect reproductive success?

A replicated laboratory experiment by Regali & Rasmont (1995) showed that four groups of captive *B. terrestris* workers fed on pollen mainly from oilseed rape *Brassica napus* ssp. *oleifera* (22% protein) reared more, larger, longer-lived males than four groups fed on pollen mainly from sunflower *Helianthus annuus* (13% protein).

Ribeiro *et al.* (1996) showed that eight laboratory-reared *B. terrestris* colonies fed on freshly frozen honey bee *Apis mellifera* pollen produced larger queens, which survived better and produced larger colonies after hibernation than queens from seven colonies fed on dried, commercially available honey bee pollen. There was no difference in the number or weight of workers or males from colonies fed on these two types of pollen.

In a replicated, controlled trial, Génissel *et al.* (2002) demonstrated that the pollen content of the diet significantly affects the fecundity of *B. terrestris* in captivity. Twenty groups of three workers fed pollen from fruit trees *Prunus* spp. or a mix of pollen including fruit tree pollen, produced more offspring (average 14–19 adult males produced/group in 95 days) than 20 groups fed pollen from dandelion *Taraxacum* sp. or willow *Salix* sp. (average 0–8 adult males produced/group). The protein content of *Prunus* pollen was shown to be higher (average 27.5%) than other pollens in the trial.

Almanza M.T. (2007) Management of *Bombus atratus* bumblebees to pollinate Lulo (*Solanum quitoense* L.), a native fruit from the Andes of Colombia. PhD thesis, University of Göttingen, Germany. www↗

Beekman M., van Stratum P. & Lingeman R. (1998) Diapause survival and post-diapause performance in bumblebee queens (*Bombus terrestris*). *Entomologia Experimentalis et Applicata*, 89, 207–214. www↗

Beekman M., van Stratum P. & Lingeman R. (2000) Artificial rearing of bumble bees *Bombus terrestris* selects against heavy queens. *Journal of Apicultural Research*, 39, 61–65. www↗

Chiang C.H., Sung I.H., Ho K.K. & Yang P.S. (2009) Colony development of two bumblebees, *Bombus eximius* and *B. sonani*, reared in captivity in a subtropical area of Taiwan (Hymenoptera, Apidae, Bombini). *Sociobiology*, 54, 699–714. www↗

Frison T.H. (1927) Experiments in rearing colonies of bumblebees (Bremidae) in artificial nests. *Biological Bulletin of the Marine Biological Laboratory, Woods Hole*, 52, 51–67. www↗

Génissel A., Aupinel P., Bressan C., Tasei J.-N. & Chevrier C. (2002) Influence of pollen origin on performance of *Bombus terrestris* micro-colonies. *Entomologia Experimentalis et Applicata*, 104, 329–336. www↗

Gosterit A. & Gurel F. (2009) Effect of different diapause regimes on survival and colony development in the bumble bee *Bombus terrestris*. *Journal of Apicultural Research and Bee World*, 48, 279–283. www↗

Gurel F. & Gosterit A. (2008) Effects of different stimulation methods on colony initiation and development of *Bombus terrestris* L. (Hymenoptera: Apidae) queens. *Applied Entomology and Zoology*, 43, 113–117. www↗

Gurel F. & Gosterit A. (2009) The suitability of native *Bombus terrestris dalmatinus* (Hymenoptera: Apidae) queen for mass rearing. *Journal of Apicultural Science*, 53, 67–73. www↗

Hasselrot T.B. (1952) A new method for starting bumblebee colonies. *Agronomy Journal*, 44, 218–219. www↗

Ings T.C., Ward N.L. & Chittka L. (2006) Can commercially imported bumble bees out-compete their native conspecifics? *Journal of Applied Ecology*, 43, 940–948. www↗

Kearns C.A. & Thomson J.D. (2001) *The natural history of bumblebees: a sourcebook for investigations*. University Press of Colorado, USA.

Kwon Y.J., Saeed S. & Duchateau M.J. (2003) Stimulation of colony initiation and colony development in *Bombus terrestris* by adding a male pupa: the influence of age and orientation. *Apidologie*, 34, 429–437. www↗

Li J., Wu J., Cai W., Peng W., An J. & Huang J. (2008) Comparison of the colony development of two native bumblebee species *Bombus ignitus* and *Bombus lucorum* as candidates for commercial pollination in China. *Journal of Apicultural Research and Bee World*, 47, 22-26. www↗

Lopez-Vaamonde C., Koning J.W., Brown R.M., Jordan W.C. & Bourke A.F.G. (2004) Social parasitism by male-producing reproductive workers in a eusocial insect. *Nature*, 430, 557–560. www↗

Plath O.E. (1923) Breeding experiments with confined *Bremus (Bombus)* queens. *Biological Bulletin of the Marine Biological Laboratory, Woods Hole*, 45, 325–341. www↗

Plowright R.C. & Jay S.C. (1966) Rearing bumble bee colonies in captivity. *Journal of Apicultural Research*, 5, 155–165. www↗

Pomeroy N. & Plowright R.C. (1980) Maintenance of bumble bee colonies in observation hives (Hymenoptera: Apidae). *The Canadian Entomologist*, 112, 321–326. www↗

Regali A. & Rasmont P. (1995) Nouvelles methods de test pour l'évaluation du regime alimentaire chez des colonies orphelines de *Bombus terrestris* (L) (Hymenoptera, Apidae) [New bioassays to evaluate diet in *Bombus terrestris* (L) (Hymenoptera, Apidae)]. *Apidologie*, 26, 273–281. www↗

Ribeiro M.F., Duchateau M.J. & Velthuis H.H.W. (1996) Comparison of the effects of two kinds of commercially available pollen on colony development and queen production in the bumble bee *Bombus terrestris* L (Hymenoptera: Apidae). *Apidologie*, 27, 133–144. www↗

Röseler P. (1985) A technique for year-round rearing of *Bombus terrestris* (Apidae, Bombini) colonies in captivity. *Apidologie*, 16, 165–170. www↗

Sladen F.W.L. (1912) *The humble bee: its life history and how to domesticate it.* Macmillan and Co., London. www↗

Tasei J-N. (1994) Effect of different narcosis procedures on initiating oviposition of pre-diapausing *Bombus terrestris* queens. *Entomologia experimentalis et applicata*, 72, 273–279. www↗

Tasei J-N. & Aupinel P. (1994) Effect of photoperiodic regimes on the oviposition of artificially overwintered *Bombus terrestris* L. queens and the production of sexuals. *Journal of Apicultural Research*, 33, 27–33. www↗

Velthuis H.H.W. & van Doorn A. (2006) A century of advances in bumblebee domestication and the economic and environmental aspects of its commercialization for pollination. *Apidologie*, 37, 421–451. www↗

Whitehorn P.R., Tinsley M.C., Brown M.J.F., Darvill B. & Goulson D. (2009) Impacts of inbreeding on bumblebee colony fitness under field conditions. *BMC Evolutionary Biology*, 9, 152. www↗

Williams P.H. & Osborne J.L. (2009) Bumblebee vulnerability and conservation worldwide. *Apidologie*, 40, 367–387.

Xerces Society (2008) *Bumblebees in decline. Invertebrate Conservation Fact Sheet.* Avail-

able at http://www.xerces.org/wp-content/uploads/2008/09/bumblebees_factsheet.
pdf. Accessed 2 December 2009.

Yeninar H., Duchateau M.J., Kaftanoglu O. & Velthuis H. (2000) Colony developmental patterns in different local populations of the Turkish bumble bee *Bombus terrestris dalmatinus*. *Journal of Apicultural Research*, 39, 107–116. www↗

Yoon H.J., Kim S.E. & Kim Y.S. (2002) Temperature and humidity favourable for colony development of indoor-reared bumblebee, *Bombus ignitus*. *Applied Entomology and Zoology*, 37, 419–423. www↗

11.2 Re-introduce laboratory-reared bumblebee queens to the wild

• We have found no evidence on the effects of reintroducing bumblebee queens.

Re-introduction of rare bumblebees to sites where they have gone extinct has not yet been tried, but a reintroduction of the short-haired bumblebee *Bombus subterraneus* is planned in southern England. In this case, inseminated queens from laboratory-reared colonies will be introduced to an area they once occupied.

11.3 Re-introduce laboratory-reared bumblebee colonies to the wild

See also Chapter 11.4 *Translocate bumblebee colonies in nest boxes.*

• Seven replicated trials have monitored the success of laboratory-reared colonies of bumblebees introduced to the environment. In four of the trials (three in the UK, one in Canada) colonies were left to develop until new queens were produced or the founding queen died. In two of these (both in the UK), the numbers of queens/colony were very low or zero. In two trials, good numbers of new queens were produced.

• Rates of social parasitism by cuckoo bees *Bombus* [Psithyrus] spp. in colonies released to the wild are variable. Two replicated trials in Canada and the UK found high rates (25–66% and 79% respectively). The UK trial showed that parasitism was reduced by placing colonies in landscapes with intermediate rather than very high nectar and pollen availability, late, rather than early in the season. Five other replicated trials reported no social parasites. We have not found evidence to compare rates of parasitism in artificial nest boxes with the rate in natural nests.

- Two replicated trials examined the effects of supplementary feeding for bumble-bee colonies placed in the field. One, in Canada, found supplementary feeding improved the reproductive success of captive-reared colonies, but did not reduce their parasite load. The other trial, in the USA, found supplementary feeding did not increase colony productivity.

- One small scale trial in Norway showed that colonies of the buff-tailed bumblebee *B. terrestris* prefer to forage more than 100 m from their nest sites.

In a replicated trial that introduced 20 commercially-reared colonies of the buff-tailed bumblebee *Bombus terrestris terrestris* into farms and 10 into suburban gardens in the UK in early June (Goulson *et al.* 2002), colonies produced an average of 160 workers. The production of new queens was variable, with averages from 21 to 36 queens/colony and no significant difference between gardens and farmland. Colonies in gardens were significantly more likely to be infested by the damaging bumblebee wax moth *Aphomia sociella* (average 77 larvae/nest, compared to 3–4 larvae/nest on farmland).

A replicated controlled trial in Canada tested the effect of feeding captive-reared bumblebee colonies (29 colonies of the common eastern bumblebee *B. impatiens*, 16 colonies of the red-belted or tricolored bumblebee *B. ternarius*) sited in a flower-rich meadow, with sucrose solution and fresh pollen (Pelletier & McNeill 2003). Twenty-one colonies that were fed produced 51% more workers, and were almost four times as likely to produce new queens as those that were not fed. Fed colonies produced between 0 and 19 queens/colony on average, and control colonies between 0 and 14 queens/colony (the paper gives separate averages for each species and each year). Social parasitism by cuckoo bees was high in this study, with between one quarter and two thirds of colonies successfully usurped. The rate was not reduced by supplementary feeding.

A trial with three laboratory-reared colonies of *B. terrestris* introduced to an agricultural landscape in Vestby, Norway (Dramstad *et al.* 2003) found that greater numbers of marked bumblebees from the colonies foraged on a 210 × 2 m sown strip of phacelia *Phacelia tanacetifolia* when the nests were moved more than 100 m away from the strip (18.3 marked bumblebees/210 m transect) than when they were placed right next to it (11.5 marked bumblebees/ 210 m transect).

Whittington & Winston (2004) placed seven laboratory-reared colonies of the rapidly declining Western bumblebee *B. occidentalis* in agricultural land and woodland in British Columbia, Canada, and compared the numbers of bees and brood with seven similar colonies prevented from foraging but supplied with water and pollen. Both groups were supplied with sugar syrup. Outside colonies produced as many workers (40–80 workers per colony on average) and more brood than enclosed colonies, but after five weeks, their syrup supply was robbed by honey bees *Apis mellifera*, resulting in high bumblebee mortality.

Ings *et al.* (2006) placed seven laboratory-reared colonies of the native UK subspecies *B. t. audax* in field locations in Surrey, UK. These colonies were left out until the founding queen died and all males and queens had emerged. They produced an average of 0.3 queens and 189 males/colony. Only two of the seven colonies produced any queens.

A replicated trial using 48 commercially-reared colonies of *B. t. terrestris* in the UK shows that rates of parasitism by cuckoo bees *Bombus* [Psithyrus] spp. can be high on colonies in nest boxes (38 colonies, 79% parasitized; Carvell *et al.* 2008). Parasitism is more intense when colonies are sited in areas of high resource availability (92% of colonies parasitized by three cuckoo bees on average among oilseed rape *Brassica napus* fields, compared to 67% parasitized by one cuckoo bee on average among wheat fields). Parasitism is also more intense if colonies are placed early in the season (early May). This suggests that if captive-reared colonies of native bumblebees are to be reintroduced, they should be placed out later in the season (early June), amongst a heterogenous landscape with intermediate levels of resource.

Whitehorn *et al.* (2009) placed 36 laboratory-reared colonies of *B. terrestris* (non-native subspecies, probably *terrestris* or *dalmatinus*) in field conditions at the University of Stirling, Scotland, from when they had 15 workers until the founding queen died. The experiment included inbred colonies and colonies with diploid males. Normal colonies (no diploid males) produced a total of 30–31 workers on average, but no new queens. Sixteen outbred colonies survived for an average of 4.5 weeks, but did not produce new queens.

A replicated trial using 19 captive-reared colonies of the long-tongued bumblebee species *B. appositus* in subalpine meadows in Colorado, USA, found that four colonies regularly fed with sugar solution did not produce significantly more workers, males or queens than 15 colonies that were not fed (Elliott 2009).

Carvell C., Rothery P., Pywell R.F. & Heard M.S. (2008) Effects of resource availability and social parasite invasion on field colonies of *Bombus terrestris*. *Ecological Entomology*, 33, 321–327. www↗

Dramstad W.E., Fry G.L.A. & Schaffer M.J. (2003) Bumblebee foraging – is closer really better? *Agriculture, Ecosystems and Environment*, 95, 349–357. www↗

Elliott S.E. (2009) Surplus nectar available for subalpine bumble bee colony growth. *Environmental Entomology*, 38, 1680–1689. www↗

Goulson D., Hughes W.O.H., Derwent L.C. & Stout J.C. (2002) Colony growth of the bumblebee, *Bombus terrestris*, in improved and conventional agricultural and suburban habitats. *Oecologia*, 130, 267–273. www↗

Ings T.C., Ward N.L. & Chittka L. (2006) Can commercially imported bumble bees out-compete their native conspecifics? *Journal of Applied Ecology*, 43, 940–948. www↗

Pelletier L. & McNeill J.N. (2003) The effect of food supplementation on reproductive success in bumblebee field colonies. *Oikos*, 103, 688–694. www↗

Whitehorn P.R., Tinsley M.C., Brown M.J.F., Darvill B. & Goulson D. (2009) Impacts of inbreeding on bumblebee colony fitness under field conditions. *BMC Evolutionary Biology*, 9, 152. www↗

Whittington R. & Winston M.L. (2004) Comparison and examination of *Bombus occidentalis* and *Bombus impatiens* (Hymenoptera: Apidae) in tomato greenhouses. *Journal of Economic Entomology*, 97, 1384–1389. www↗

11.4 Translocate bumblebee colonies in nest boxes

- We have captured three small trials in the 1950s and early 1960s testing the effect of translocating bumblebee colonies in nest boxes. Two trials in Canada provided evidence of queen death and one of these showed lower colony productivity following translocation. Just one, a UK trial, concluded that early bumblebee *Bombus pratorum* colonies adapt well to being moved.

An unspecified number of red-belted bumblebee *Bombus rufocinctus* colonies in wooden nest boxes were translocated an unspecified distance from their original site to a crop field, in southern Alberta, Canada, once the first brood of workers had begun foraging (Hobbs *et al.* 1960). Some workers were lost and queens began foraging for nectar. Two queens were killed as a result of returning to the wrong nest. Colonies that were moved raised an average of four new queen cocoons (range 4–9), while colonies that were not moved raised on average 22 new queen cocoons (range 17–27).

Five colonies of the early bumblebee *B. pratorum* housed in wooden boxes were experimentally translocated in Hertfordshire, southern UK (Free 1955). Colonies were moved six miles (9.7 km), 80 yards (73 m), seven yards (6.4 m) or three inches (7.6 cm) from the original site. Apart from the smallest translocation distance, some foraging workers were lost due to each translocation. Between 71% and 92% of foraging workers returned to the nest at the new site eventually.

On seven occasions in spring 1960 and 1961, an unspecified number of colonies of long-tongued bumblebee species *B. appositus*, *B. californicus* and *B. nevadensis* were moved, overnight, up to six miles away just after the first brood of workers had emerged, in southern Alberta, Canada (Hobbs *et al.* 1962). An empty box with a one-way door was placed on the old site for two hours the following morning. On one occasion, half the workers from a colony of *B. californicus* were out when the colony was moved and captured in the trap. On the other six occasions, few workers were left behind. Several queens (at least eight) were killed after translocation by re-entering the wrong nest box. The authors recommend delaying translocation until the second brood has emerged and the queen no longer forages.

Free J.B. (1955) The adaptability of bumblebees to a change in the location of their nest. *British Journal of Animal Behaviour*, 3, 61–65. www↗

Hobbs G.A., Virostek J.F. & Nummi W.O. (1960) Establishment of *Bombus* spp. (Hymenoptera: Apidae) in artificial domiciles in Southern Alberta. *The Canadian Entomologist*, 92, 868–872. www↗

Hobbs G.A., Nummi W.O. & Virostek J.F. (1962) Managing colonies of bumble bees (Hymenoptera: Apidae) for pollination purposes. *The Canadian Entomologist*, 94, 1121–1132. www↗

11.5 Rear and manage populations of solitary bees

See also Chapter 10 *Providing artificial nest sites for solitary bees* and Chapter 11.6 *Translocate solitary bees*.

• Several species of solitary bee in the family Megachilidae are reared and managed commercially as pollinators, mostly for the forage crop alfalfa, or fruit trees. These species readily nest in drilled wooden blocks, or stacked grooved boards of wood or polystyrene. Parasites and pathogens can be problematic and a number of control methods have been developed. Rearing methods have been investigated for two other species not yet commercially managed and one replicated trial shows that temperature regimes are important to survival. If rearing for conservation purposes is to be attempted, we would recommend a systematic review of these methods.

- Three management trials with megachilids not commercially managed in the USA or Poland, and a review of studies of managed species, found that local populations can increase up to six-fold in one year under management if conditions are good and plentiful floral resources are provided.

- Two replicated trials have reared solitary bees on artificial diets. One found high larval mortality in *Osmia cornuta* reared on artificial pollen-based diets, including honey bee-collected pollen. The other found *Megachile rotundata* could be reared on an artificial diet based on honey bee-collected pollen, but bees reared on synthetic pollen substitutes either died or had low pre-pupal weight.

Background

Techniques for rearing solitary bees of the family Megachilidae have been developed primarily for pollination purposes and there is a substantial literature on husbandry techniques for species such as the alfalfa leafcutter bee *Megachile rotundata*, and the mason bees *Osmia cornuta*, *O. cornifrons* and *O. lignaria* (reviewed by Bohart 1972, Torchio 1987, Richards 1993, Bosch & Kemp 2002).

The ITIS world bee checklist lists 4,048 species in the Megachilini and some are declining or threatened in some areas. For example, there are four species of mining bee *Osmia* on the UK Biodiversity Action Plan. Captive rearing may be one strategy to augment or re-establish their populations. We acknowledge that the rare or declining species are likely to differ biologically and ecologically from the species selected for management. But we have come across some evidence on rearing techniques for Megachilid bees that seems relevant in the context of captive rearing for conservation purposes, either because it highlights where there are differences between species, or it tests rearing methods for currently unmanaged species.

To provide an overview of rearing techniques, we summarise the most recent review (Bosch & Kemp 2002) below. If captive rearing becomes a popular conservation strategy for solitary bees, we would recommend a systematic review of these methods.

Rearing methods for solitary bees

Bosch & Kemp (2002) review methods that have been developed for rearing three species of mason bee now used as orchard pollinators in Japan, USA and Europe respectively: *Osmia cornifrons*, *O. lignaria*, and *O. cornuta*. All three species will nest in holes drilled in wood or polystyrene, grooved wood or polystyrene boards stacked together, paper or cardboard tubes or reed stem sections. If nest cavities are too narrow or too short, more males will be reared. Temperature regimes are important to survival through four of the seven developmental stages identified: development (egg to adult, including dormant pre-pupal phase), pre-wintering, wintering and incubation prior to emergence. Responses to temperatures differ between species, and between populations from different areas within species. These should be experimentally studied to develop an effective rearing regime.

Releasing bees at a site in their nests (rather than as extracted cocoons) increases the chance of females nesting at the same site, but extracting cocoons is used to reduce the spread of fungal pathogens in the alfalfa leafcutter bee *Megachile rotundata*.

Exposure to predators and parasites can be reduced by: covering nest shelters with a screen (deters birds); releasing bees in excluder boxes to avoid re-use of old nests (fungus/mites); removing nest boxes after nesting activity (late-flying parasitoids); black light traps (kills *Monodontomerus* and other wasp enemies); acaricide and thermal shock treatment during dormant pre-pupa stage (reduces mite numbers); selective trapping of wasp predators (e.g. male *Sapyga pumila* wasps can be caught at night in 2.5 mm diameter cavities).

One replicated trial in India examined methods of artificially rearing the subtropical leafcutter species *Megachile flavipes*, a species not commercially managed for pollination (Kapil & Sihag 1985). This species can be stored at the pre-pupal stage for 270 days. It showed least mortality (5–7%) under this treatment if stored at 12°C. Stored at 4°C, the lowest temperature in the study, mortality was 16–20%. The optimum incubation temperature following lower temperature storage was 28.5°C. The bees emerged over a shorter time period if stored at 8°C and incubated at 28.5°C.

A trial with the sunflower leafcutter bee *Megachile pugnata* in Utah, USA, found that females will nest in drilled wooden nest blocks, preferring holes 15 cm deep (Parker & Frohlich 1985). They can be overwintered from November to June or July at 3°C in the laboratory, and emerge after incubation at 30°C. On release in a sunflower field, female bees of the species were recorded foraging and nesting in the field where they were released (released individuals not marked).

Native Australian bees of the genera *Amegilla* and *Xylocopa* are being considered as potential managed pollinators (see for example Hogendoorn *et al.* 2007), but we have found no evidence on the efficacy of captive rearing techniques for these bees.

Can populations be augmented by rearing?

A trial with *Osmia sanrafaelae*, native to the San Rafael Desert, Utah, USA, found that bees were induced to nest in pine wood nest boxes with drilled 9 mm holes inside a 6 × 6 × 2 m saran cloth cage placed over an alfalfa crop *Medicago sativa* (Parker 1985). Fifty males and 50 females were introduced to the cage in July, and although mortality in the nests was high (47%) the number of bees surviving to adulthood in the next generation was 4-fold higher than the number originally introduced (exact number not given).

In a trial at two experimental farms near Poznan, western Poland from 1989 to 1994, the numbers of red mason bees *Osmia rufa* nesting in bundles of reed stem increased substantially year on year. Each winter, occupied reed stems were collected and healthy bee cocoons (not parasitized) were transferred to refrigerators and kept at 4°C over winter. These were placed out in incubators along with new nest boxes the following spring. At one site, the number of emerging bees increased from an originally introduced 1,453 bees in 1989 to 108,973 in 1994 (a 75-fold increase; Wójtowski *et al.* 1995). The number of emerging females each year was between 1.3 and 5.7 times the number of females the previous year. Based on these numbers, the density of red mason bees on the farm was estimated to have increased from 1 bee/ha to 1,353 female bees/ha or more over the six years. Bee numbers nesting at the second site followed a similar trajectory, but the experiment was ended after three years.

A review of captive-rearing methods developed for orchard bees in the genus *Osmia* reports evidence that female populations have been increased by 2- to 3-fold for *O. cornifrons* and 5-fold for *O. lignaria* in orchards, in years with good weather and fruit tree flowering (Bosch & Kemp 2002). Poor weather during flowering or short blooming periods can lead to population losses (no experimental evidence reported).

Captive-reared sunflower leafcutter bees *Megachile pugnata* were released into a 0.7 ha sunflower field in Utah, USA in 1982 (Parker & Frohlich 1985). A total of 186 females were released (not marked) and a maximum of 182 were subsequently counted resting in nesting blocks at night, in and around the field. Altogether, 690 nests were made across 90 nest blocks, and the surviving number of overwintering pupae when counted in October was 1,643.

Can solitary bees be reared on artificial diets?

A replicated controlled laboratory experiment at the University of Bologna, Italy, found that 408 European orchard bees *Osmia cornuta* reared on pollen-based artificial diets showed high larval mortality (76–100%; Ladurner *et al.* 1999). The 331 control bees allowed to consume the pollen lump provided by their mother had lower mortality rates (4–32%). These authors suggest this is because female *O. cornuta* bees add something, perhaps an enzyme, to the pollen they provide for their larvae.

Nelson *et al.* (1972) reared groups of up to 25 alfalfa leafcutter bees *Megachile rotundata* on two different pollen substitutes or honey bee-collected pollen, and compared them to larvae reared on pollen collected by the mother bee. Those reared on pollen substitutes either died (one type of synthetic diet) or had lower prepupal weights (average weights 15–40 mg) than the control group (55 mg). Bees reared on honey-bee collected pollen weighed more than the control group (average weights 64–71 mg).

Bohart G.E. (1972) Management of wild bees for the pollination of crops. *Annual Review of Entomology*, 17, 287–312.

Bosch J. & Kemp W.P. (2002) Developing and establishing bee species as crop pollinators: the example of *Osmia* spp. (Hymenoptera: Megachilidae) and fruit trees. *Bulletin of Entomological Research*, 92, 3–16. www↗

Hogendoorn K., Coventry S.A. & Keller M.A. (2007) Foraging behaviour of a blue banded bee, *Amegilla (Notomegilla) chlorocyanea* Cockerell in greenhouses: implications for use as tomato pollinators. *Apidologie*, 38, 86–92.

Kapil R.P. & Sihag R.C. (1985) Storage and incubation in the management of the alfalfa-pollinating bee *Megachile flavipes* Spinola. *Journal of Apicultural Research*, 24, 199–202. www↗

Ladurner E., Maccagnani B., Tesoriero D., Nepi M. & Feliciolo A. (1999) Laboratory rearing of *Osmia cornuta* Latreille (Hymenoptera Megachilidae) on artificial diet. *Bollettino* dell'Istituto di Entomologia della Università di Bologna, 53, 133–146. www↗

Nelson E.V., Roberts R.B. & Stephen W.P. (1972) Rearing larvae of the leaf-cutter bee *Megachile rotundata* on artificial diets. *Journal of Apicultural Research*, 11, 153–156. www↗

Parker F.D. (1985) A candidate legume pollinator, *Osmia sanrafaelae* Parker (Hymenoptera: Megachilidae). *Journal of Apicultural Research*, 24, 132–136. www↗

Parker F.D. & Frohlich D.R. (1985) Studies on management of the sunflower leafcutter bee *Eumegachile pugnata* (Say) (Hymenoptera: Megachilidae). *Journal of Apicultural Research*, 24, 125–131. www↗

Richards K.W. (1993) Non-*Apis* bees as crop pollinators. *Revue Suisse de Zoologie*, 100, 807–822.

Torchio P.F. (1987) Use of non-honey bee species as pollinators of crops. *Proceedings of the Entomological Society of Ontario*, 118, 111–124.

Wójtowski F., Wilaniec Z. & Szymaś B. (1995) Increasing the total number of *Osmia rufa* (L.) (Megachilidae) in selected biotopes by controlled introduction method. Pages 177–180 in: Ed, Banaszak J. *Changes in the fauna of wild bees in Europe*. Pedagogical University, Bydgoszcz. www↗

11.6 Translocate solitary bees

See Chapter 11.5 *Rear and manage populations of solitary bees* for evidence on establishing and augmenting managed solitary bee populations.

• One replicated trial in India showed that translocating carpenter bees *Xylocopa fenestrata* in immature stages can establish a population at a new site, but if adult bees are translocated, a very small proportion remain at the new site.

A study of the subtropical carpenter bee *Xylocopa fenestrata* on agricultural land in Haryana, India, found that populations could be translocated to a site 3 km away, if this was done with immature stages sealed within hollow stems (Sihag 1993). Of 90 translocated adult bees, only three remained at the new site. Around 40% of bees translocated as young (90 stems moved with young bees inside) stayed after emergence at the new site; 63–70% of these translocated females stayed and began provisioning nests, whereas most males left the site after territorial fights.

Sihag R.C. (1993) Behaviour and ecology of the subtropical bee *Xylocopa fenestrata* F. 7. Nest preferences and response to nest translocation. *Journal of Apicultural Research*, 32, 102–108. www↗

11.7 Introduce mated females to small populations to improve genetic diversity

- One trial in Brazil showed that genetic diversity can be maintained in small isolated populations of social bees by regularly introducing inseminated queens.

Background

Bees are vulnerable to a particular type of extinction vortex brought about by the occurrence of sterile diploid males when genetic diversity is low (Zayed & Packer 2005). It happens because bees (and other insects in the Hymenoptera) develop into females when there are two different forms (alleles) of the sex-determining gene present (heterozygous). Males are normally haploid, having only one set of chromosomes not two, so heterozygosity is impossible. With low genetic diversity, diploid individuals that would normally be females can end up with identical alleles and develop into males, called diploid males. They are generally sterile, or short-lived. In stingless bees *Melipona* spp., diploid males and the queen that produced them are killed by the workers, which results in colony failure (Carvalho 2001).

The risk of a diploid male extinction vortex may be high in rare species, or those living in fragmented populations. For example, up to 32% of males were diploid in some species of Euglossine bees in Colombia (Lopez-Uribe *et al.* 2007), and three of 16 populations of the rare bumblebee *Bombus muscorum* in the UK had diploid males (Darvill *et al.* 2006).

Carvalho (2001) established a small isolated population of the Brazilian stingless bee *Melipona scutellaris*, based on 22 wild-collected colonies. She introduced between three and 13 inseminated queens each year over four consecutive years, and found that the small population retained diversity in its sex allele over nine years. It did not increase its production of diploid males or collapse to extinction, as might have been expected. Carvalho recommends exchange of inseminated queens between beekeepers as a way to ensure the survival of small meliponiaries.

Carvalho G.A. (2001) The number of sex alleles in a bee population and its practical importance. *Journal of Hymenoptera Research*, 10, 10–15. www↗

Darvill B., Ellis J.S., Lye G.C. & Goulson D. (2006) Population structure and inbreeding in a rare and declining bumblebee, *Bombus muscorum* (Hymenoptera : Apidae). *Molecular Ecology*, 15, 601–611.

Lopez-Uribe M.M., Almanza M.T. & Ordonez M. (2007) Diploid male frequencies in Colombian populations of euglossine bees. *Biotropica*, 39, 660–662.

Zayed A. & Packer L. (2005). Complementary sex determination substantially increases extinction proneness of haplodiploid populations. *Proceedings of the National Academy of Sciences of the United States of America*, 102, 10742–10746.

12. Education and awareness-raising

Enhance bee taxonomy skills through higher education and training

We have captured no evidence for the effects of developing taxonomy skills.

Provide training to conservationists and land managers on bee ecology and conservation

We have captured no evidence for the effects of providing training.

Raise awareness amongst the general public through campaigns and public information

We have captured no evidence for the effects of such campaigning techniques.

Background

There is a strong need for awareness-raising, education and training about the diversity of wild bees, their conservation and the services they provide (Brown & Paxton 2009). The International Pollinator Initiative (IPI) of the United Nations Convention on Biological Diversity has awareness-raising amongst scientists, policymakers and the public as one of its central aims (Byrne & Fitzpatrick 2009).

Brown M.J.F. & Paxton R.J. (2009) The conservation of bees: a global perspective. *Apidologie*, 40, 410–416.

Byrne A. & Fitzpatrick U. (2009) Bee conservation policy at the global, regional and national levels. *Apidologie*, 40, 193–210.

12.1 Enhance bee taxonomy skills through higher education and training

- We have captured no evidence for the effects of developing taxonomy skills on bee conservation.

12.2 Provide training to conservationists and land managers on bee ecology and conservation

See Chapter 6.2 *Replace honey-hunting with apiculture* for evidence relating to a training programme for stingless beekeepers.

- We have captured no evidence for the effects of providing training on bee ecology and conservation to conservationists and land managers.

12.3 Raise awareness amongst the general public through campaigns and public information

- We have captured no evidence for the effects of campaigning or raising awareness about bees and their conservation.

There has been a large amount of campaigning to the public and to policymakers about bee conservation in recent years, particularly in Europe and North America. It is possible that this awareness-raising has already resulted in direct benefits for bees, in the management of urban and rural landscapes. However, we have found no studies examining the effects of awareness-raising in changing the way people behave or the way land is managed.

Index

About Pelagic Publishing

We publish books for scientists, conservationists, ecologists, wildlife enthusiasts – anyone with a passion for understanding and exploring the natural world. Working closely with authors and organisations we are publishing books that:

- Deliver cutting-edge knowledge, published rapidly in traditional and eBook formats.
- Promote best-practise in research techniques and management methods.
- Encourage and assist practical wildlife investigation and field exploration.
- Bridge the gap between scientific theory and practical implementation.
- Share wildlife experiences.
- Promote taxonomic and identification skills.
- Highlight the use of technology in science and wildlife exploration.

Stay up-to-date with news and offers - visit www.pelagicpublishing.com

A discount on your next book

Get 15% off your next order from Pelagic Publishing by quoting 'BCRB15' when you order from our website (www.pelagicpublishing.com). Terms and conditions apply: this coupon can only be used once per customer and cannot be used in conjunction with any other offer.